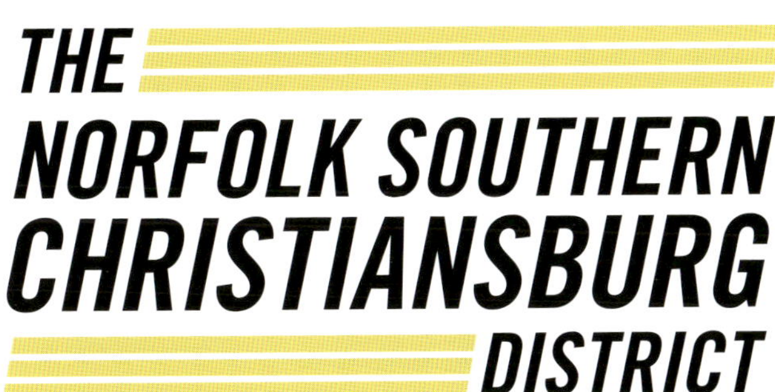
THE
NORFOLK SOUTHERN
CHRISTIANSBURG
DISTRICT

# THE NORFOLK SOUTHERN CHRISTIANSBURG DISTRICT

## A MODERN MOUNTAIN RAILROAD

DANIEL ALLS

AMERICA THROUGH TIME®
ADDING COLOR TO AMERICAN HISTORY

America Through Time is an imprint of Fonthill Media LLC
www.through-time.com
office@through-time.com

Published by Arcadia Publishing by arrangement with Fonthill Media LLC
For all general information, please contact Arcadia Publishing:
Telephone: 843-853-2070
Fax: 843-853-0044
E-mail: sales@arcadiapublishing.com
For customer service and orders:
Toll-Free 1-888-313-2665

www.arcadiapublishing.com

First published 2023



ISBN 978-1-63499-452-1



Typeset in 10pt on13pt Sabon
Printed and bound in England

# Introduction

The Norfolk Southern Christiansburg district spans right at 100 miles and runs through the mountains of Virginia and West Virginia. The district begins just outside of Roanoke, Virginia's North Yard, and ends at the yard in Bluefield, West Virginia. In between Roanoke and Bluefield are beautiful rolling mountains, rivers, tunnels, trestles, and many great photo locations.

A portion of the Christiansburg district began its life as the Virginia and Tennessee Railroad. The railroad was built in the 1850s and ran from Lynchburg, Virginia, west to Bristol, Tennessee. The railroad's official completion date was October 1, 1856. The railroad was used to transport freight and was also used for troop transport for the Confederate army. The railroad eventually became part of the Atlantic, Mississippi and Ohio Railroad (AM&O) and was owned by William Mahone of Lynchburg, Virginia. In 1881, the AM&O was purchased and was renamed the Norfolk and Western.

Norfolk and Western then extended the line and also branched off in Walton, Virginia, to expand their operations up into the coal fields of West Virginia. The N&W operated from 1838 to 1982 and was formed by over 200 railroads merging into one. The Norfolk and Western was known as "King Coal" and a major part of its coal route ran between the coal fields of West Virginia and down onto the current day Christiansburg district.

In 1982, the Norfolk and Western Railway and Southern Railway merged to form the modern-day Norfolk Southern. The line ran from a bustling North Yard in Roanoke, Virginia, west to an also very busy Bluefield, West Virginia, yard and became known as the Christiansburg District. The district's namesake town, Christiansburg, is located about 27 miles west of Roanoke and is the crest of two major grades on the district. The grade from Elliston, Virginia, up to the crest at Christiansburg is a steady 1.5 percent. From Christiansburg down to around Walton, Virginia, is known as the Charleston grade. The grade is lesser at 1 percent, but still grueling for loaded eastbound trains out of Bluefield.

There are many Norfolk and Western relics along the way throughout the district that are still standing including many old whistle posts and a couple coaling towers.

The Norfolk and Western steam giants may not roam the rails anymore, but now there are many high horse-powered diesel models that keep America's goods on the move. In this book, we will see all of the current models from EMD and GE. EMD stands for Electro-Motive Division, which is a division of General Motors. The competing manufacturer is GE, which stands for General Electric. EMD's current mainline models are the SD70ACe, the SD70ACU, and the SD70ACC. The current GE models are the Dash 9, AC44C6M, ET44AC, ES44DC, and the ES44AC.

We will see each of these Norfolk Southern models traversing the entire Christiansburg district along with power from other railroads around the country. Some of the common "foreign power" around here are Burlington Northern Santa Fe, Union Pacific, Canadian National, Canadian Pacific, and CSX. They will be pulling manifests (freight trains), loaded and empty coal hopper trains, grain trains, intermodals, ethanol trains, and more. There will also be at least one photo of each of Norfolk Southern's twenty heritage units. Each of these heritage units is painted in schemes of predecessor railroads, which include paint schemes from the Norfolk and Western Railway and Southern Railway. There are ten EMD SD70ACe heritage units and ten GE ES44AC schemes. The twenty heritage units debuted at a special event at the North Carolina Transportation Museum in Spencer, North Carolina, in 2012. These units offer a nice splash of color against the otherwise black and white paint scheme that Norfolk Southern uses. We will also see engines being used as "DPUs," which stands for distributive power units. These are unmanned units placed in the middle or the rear of trains. They are controlled by the lead unit and are on most trains at this time.

Not every location will be covered in the next 200 photos, but most will be. We will start just outside of Roanoke, Virginia, in Roanoke County. From there, we will meander our way west into Montgomery County and Giles County, Virginia, before crossing into Mercer County, West Virginia, where the district ends in Bluefield Yard. There will also be the history of some of the locomotives pictured along with some history of the towns that the trains pass through.

I would like to thank some very special people for helping me along the way in this hobby. First off, thank you to Haley, who has been by my side for over seven years at the time of writing this. She has gone on many adventures with me chasing trains across many states.

I would also like to thank my close railfanning friends who I have met along the way: Mike P. for helping me so much with historical information for the book and for being the first person to really help me gain the knowledge I have today about the railroading world; and Chris E. for our many trips along the east coast the last few years to see many different railroads and the many days along the entire Christiansburg district.

Next I would like to thank my "West Virginia" friends Michael W. and Joseph H. There have been countless impromptu trips around the area, and most are started with texts or phone calls about something decent coming through the area. I would also like to thank my friend Sam P. for all the trips to some of the most obscure and remote places for train photos and late nights flashing trains, sometimes until well after three in the morning.

# The Norfolk Southern Christiansburg District

The journey westward across the Christiansburg district begins just outside of Roanoke, Virginia's North Yard. Here we see westbound hopper train NS 821 as it clears control point WB, which is the first control point out of the yard. From here, the train has approximately 100 miles before it rolls into Bluefield yard in West Virginia. WB is milepost N262.9.

Eastbound autorack train NS 272 has control point WB in front of it as it heads toward North Yard before heading down the Winston Salem district to its final destination of Winston Salem, North Carolina. The autoracks are loaded with vehicles built and assembled in and around Detroit, Michigan.

EMD SD70ACe 1129 leads manifest NS 126 east through Salem, Virginia, on an overcast morning. The train is only a few miles out of Roanoke's North Yard, where it will terminate. No. 1129 was built in November 2014.

An eastbound manifest rumbles east through Salem with an AC44C6M rebuild on point. This view is from atop the Lee Highway bridge. No. 4039 is ex-standard cab Dash 9 no. 8810.

Intermodal train NS 29G has left Roanoke's North Yard and is now well into the Christiansburg district as it rolls through a foggy Salem, Virginia. An ES44AC leads the long train. ES44ACs are numbered from 8000 to 8184 and were all built between 2008 and 2017. Ten of these units are painted in special heritage schemes.

A General Electric trio, with the leader being a Dash 9, pulls intermodal NS 23G east through Salem, VA. It is seen passing through an intermediate signal as it is about to pound over the McClelland Street crossing. Soon, Dash 9s will be a thing of the past with them being rebuilt into AC44C6Ms.

*Opposite above:* Seen from the same location as the last photo, but with less foggy conditions, manifest NS 353 rolls west with an ES44DC leader. The ES44DCs are numbered from 7500 to 7719 and were built between 2004 and 2008. These units were originally built with 4,000 horsepower, but were all later upgraded to 4,400 horsepower.

*Opposite below:* Looking in the opposite direction from the last photo, manifest NS 126 rushes east through Salem as it approaches the Union Street crossing. An ES44DC leads the train toward its termination at Roanoke's North Yard.

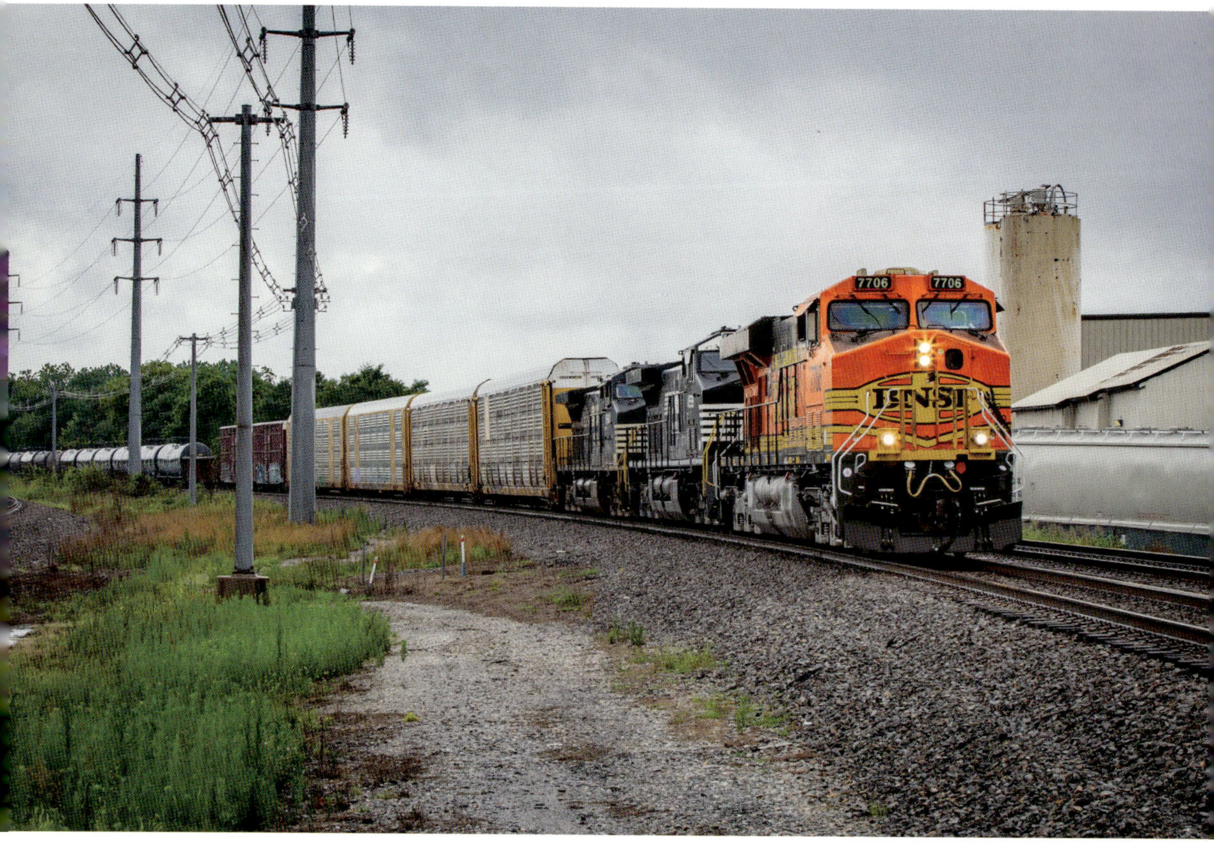
7706
7706
BNSF

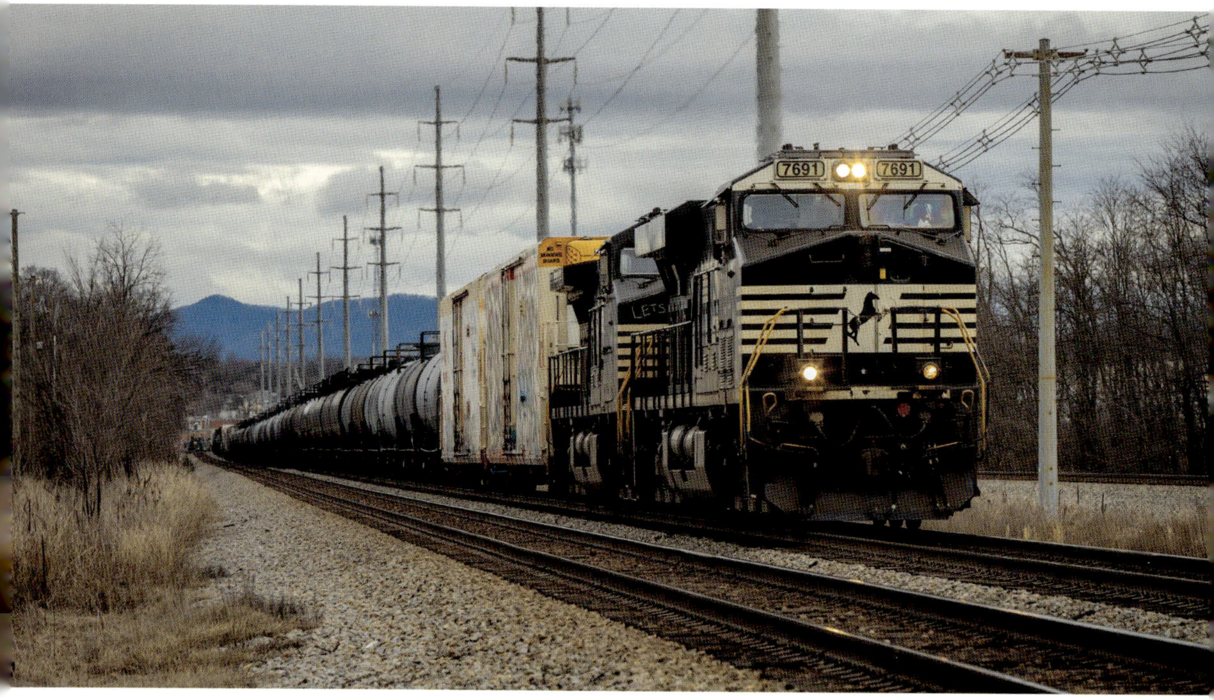
7691
7691

One of Norfolk Southern's twenty specially painted heritage units, the Nickel Plate Road heritage unit, leads hot intermodal NS 218 east through control point VN just as the sun comes up for the day. Control point VN is milepost 267.3. NS 218 operates from Calumet Yard in Chicago, Illinois, to Satelite Yard in Charlotte, North Carolina.

*Opposite above:* A Burlington Northern Santa Fe ES44DC in Heritage 2 (H2) paint is on the head end of manifest NS 18M as it powers east through Salem. Just ahead of it is the very busy Mill Lane crossing. 18M is a daily manifest that operates from Elkhart, Indiana, to Crewe, Virginia.

*Opposite below:* An ES44DC leads an empty ethanol train west through Salem, Virginia, as it approaches control point VN. In this section of Salem, the Christiansburg district parallels the Whitethorne district as seen in the photo. The Whitethorne breaks off from the Christiansburg in Narrows, VA, and primarily is used for heavy eastbound traffic from Bluefield due to it having an easier grade.

The signals of control point VN and Glenvar on the Christiansburg district and the Whitethorne district can be seen behind empty grain train NS 41N as it rushes west. A "white face" Dash 9 leads the long empty train. The white face refers to the extra white paint around the windshields. This was an experimental paint scheme only placed onto approximately seventeen Dash 9s.

With an SD70ACe in the lead, manifest NS 126 rushes east through control point Glenvar and is about to cross the Diuguids Lane crossing on a snowy evening. Glenvar is milepost 267.7.

A pair of old logo Dash 9s lead hopper train NS 815 west through the industrial area of Glenvar, Virginia. At this time, the Dash 9s were going through an extensive rebuild process into AC44C6Ms and finding a train with only Dash 9s for power, especially old logo units, was very hard. The "old logo" refers to them not having the new NS logo with the thoroughbred on the side.

The clouds begin to thin out at just the right time for a decent sunset as intermodal NS 202 rushes east toward Roanoke through Glenvar, VA. The Jersey Central Line's heritage EMD SD70ACe leads the train along with another ACe. No. 202 is a daily intermodal that operates from Rossville, Tennessee, to Allentown, Pennsylvania.

With a single rusty-nosed ES44DC serving as rear DPU, eastbound intermodal NS 23G rushes through Glenvar, VA, just past the Porterfield intermediate signal. It has also just crossed over the Poor Mountain Road crossing.

Eastbound intermodal NS 234 sweeps the curve just past the Porterfield intermediate signal with special paint scheme AC44C6M no. 4004 in the lead. The blue stripes, or "mane," signify that this AC44C6M was a rebuild completed in the Juniata Locomotive shops in Altoona, Pennsylvania. No. 4004 is former standard cab Dash 9 no. 8866. This train has just passed under the "Porterfield" intermediate signal just out of frame.

Now we begin moving into what the Christiansburg District is famous for—mountains. Here in Wabun, Virginia, manifest NS 353, with an EMD SD70ACe leader, heads west on a beautiful early fall morning. Wabun offers many photo opportunities for any time of the day.

The orange Interstate heritage unit speeds west through Wabun with intermodal NS 22A on a cold and clear winter morning. NS 22A is a daily intermodal that operates from Norfolk, Virginia, to Louisville, Kentucky.

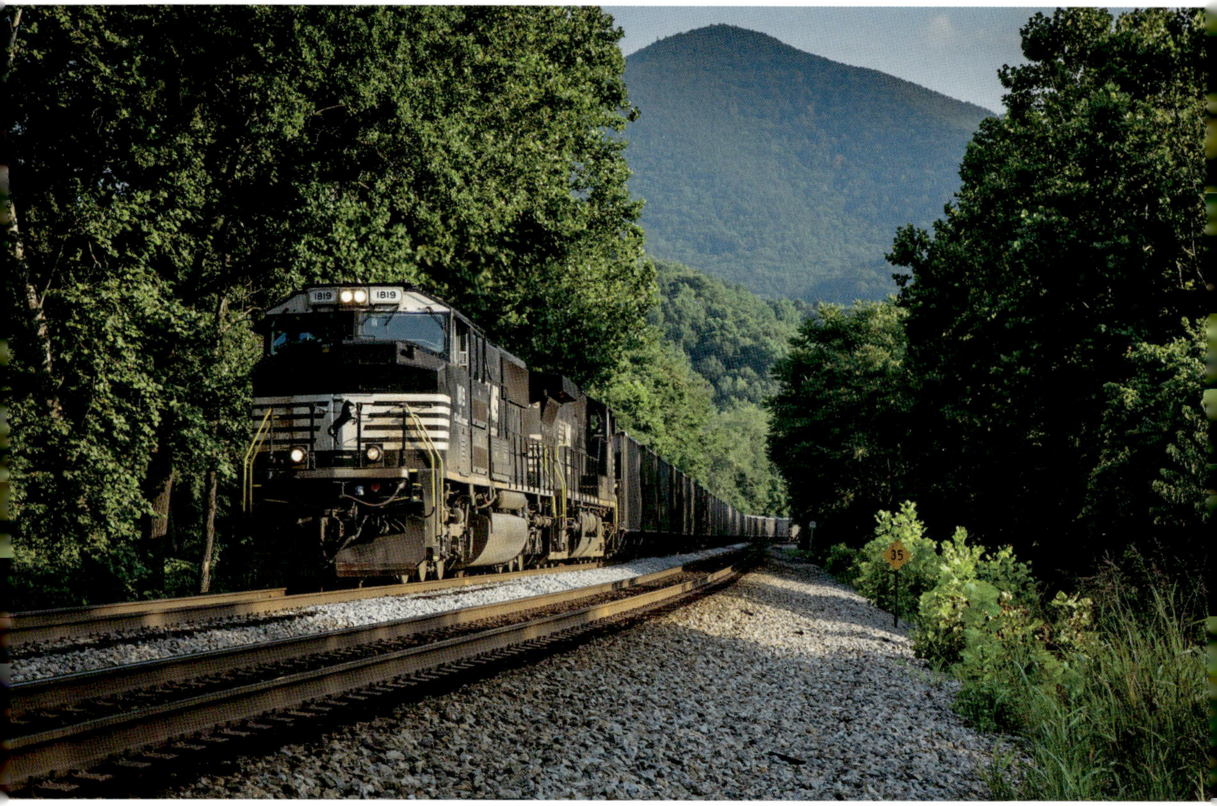
1819
1819
35

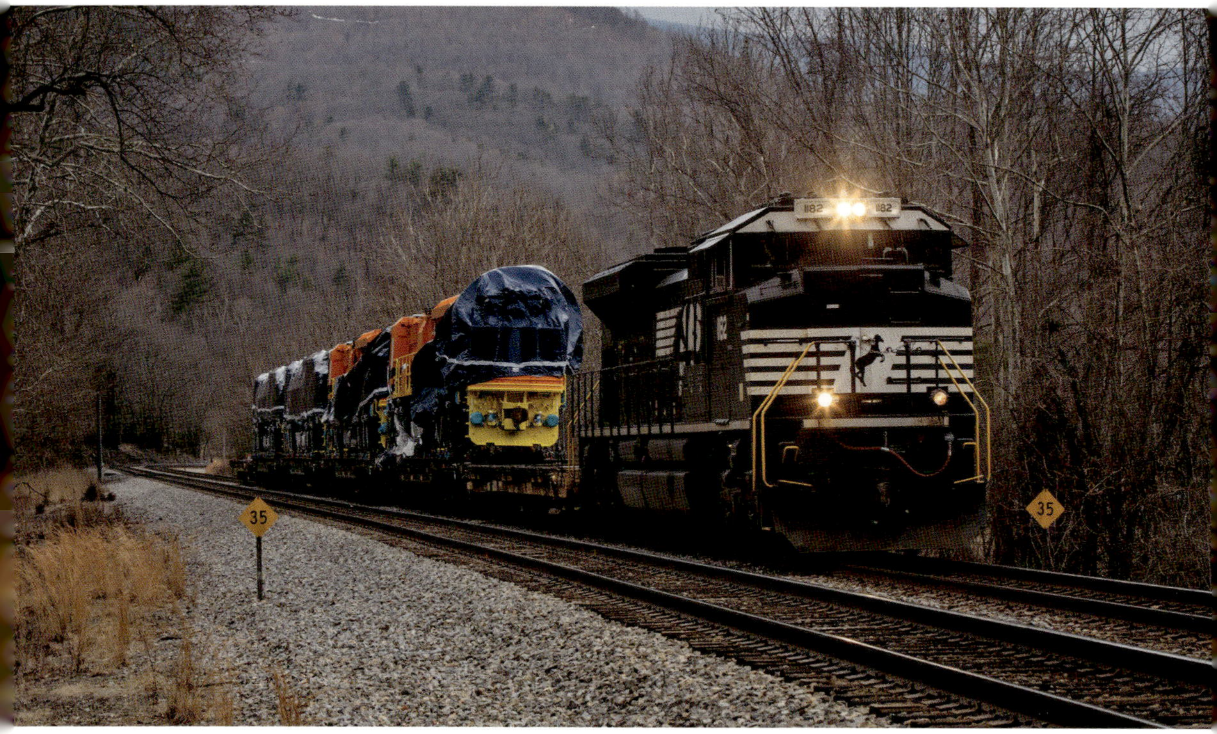
1182
1182
35
35

Rolling west in Wabun is intermodal NS 29G with a trio of old logo Dash 9s for power.

*Opposite above:* Loaded rock train NS 68Q rumbles east through Wabun with an SD70ACC rebuild showing the way. This train was loaded at a quarry in Princeton, WV, off of the Princeton Deepwater District. SD70ACC no. 1819 is former SD70 no. 2546.

*Opposite below:* Export locomotives head east toward Norfolk, Virginia. They are seen here in Wabun, Virginia, with an EMD SD70ACe leader on an overcast winter morning.

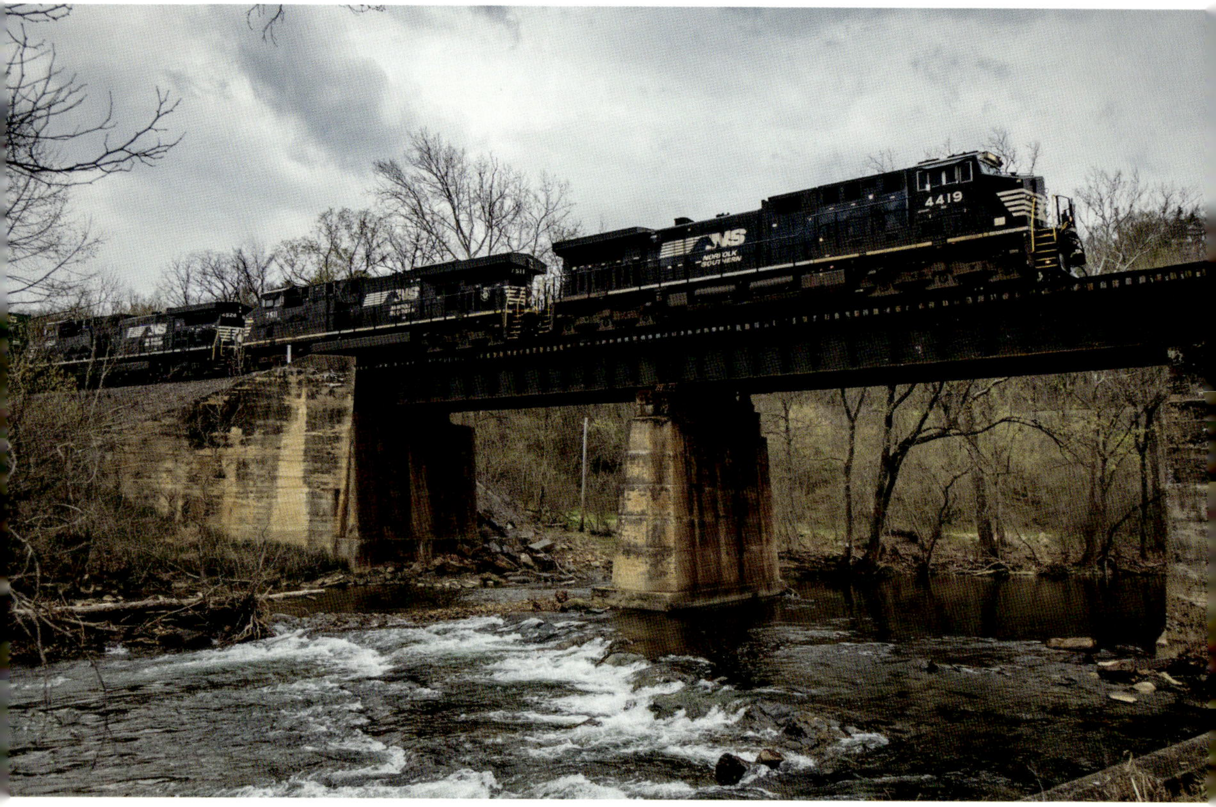
4419
NS
NORFOLK SOUTHERN

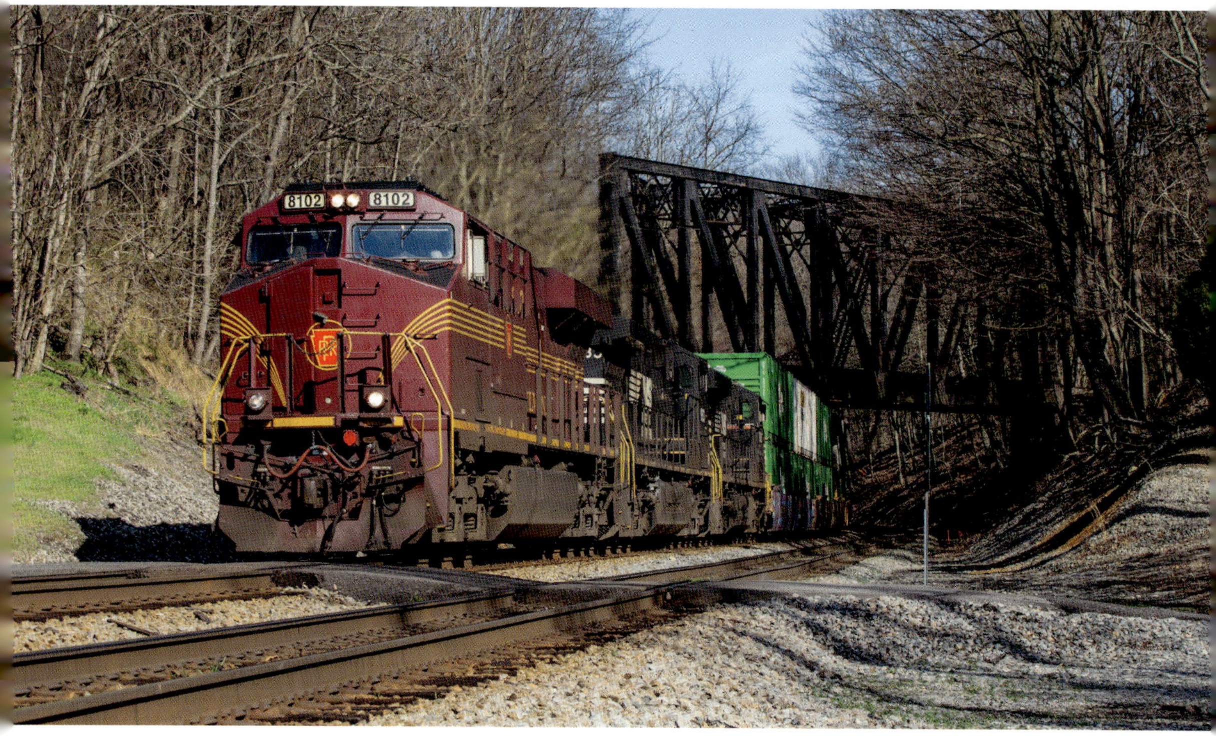
8102
8102

The sound of a loud horn echoes around the area well before westbound empty potash train NS 61P comes into view. A bright red Canadian National SD70M-2 leads the train as it throttles up and swings a curve just short of control point Singer in Wabun, Virginia.

*Opposite above:* Intermodal NS 201 heads west over the Roanoke River in Wabun, Virginia. No. 201 is a daily intermodal that operates from Allentown, Pennsylvania, to Rossville, Tennessee. Unit 4419 is ex-Dash 9 no. 9176 and was rebuilt in January 2021.

*Opposite below:* The Pennsylvania Railroad heritage unit rushes west through Wabun, Virginia, with intermodal NS 217. Above the train is the Norfolk Southern Whitethorne district. That line runs from Narrows, Virginia, where it meets the Christiansburg district and terminates in Roanoke, Virginia. The Whitethorne, ex-Virginian, handles mostly eastbound traffic out of Bluefield due to the lesser grades on the line for the heavy loaded trains.

As we move our way west on the line, we come up to the curve that leads to control point Singer. This control point is located in Wabun, Virginia, and is the last location in Roanoke County, just to the west is Montgomery County. Here an ES44AC with black number boards leads intermodal 29G on a cold winter afternoon.

A Canadian National GE is in the lead of loaded ethanol train NS 6W4 as it swings the curve at mile post N273 in Wabun, VA, next to control point Singer. Control point Singer is milepost 273.3, as seen on the concrete milepost sign.

A big SD70ACU shoves on the rear of a loaded domestic coal train as it grinds toward Roanoke. It is seen here passing the mile marker at control point Singer in Wabun. SD70ACUs were rebuilt from EMD SD90MACs and are numbered from 7229 to 7339. At this time, most of the active ACUs are in helper service on the Pittsburgh Line in Pennsylvania. Many others have been sold mainly for scrap.

A heritage duo of the Lehigh Valley and Nickel Plate Road heritage units roll through the signals of control point Singer in Wabun, Virginia. The Lehigh Valley operated from 1846 to 1976 and was originally formed to transport anthracite through New York and Pennsylvania.

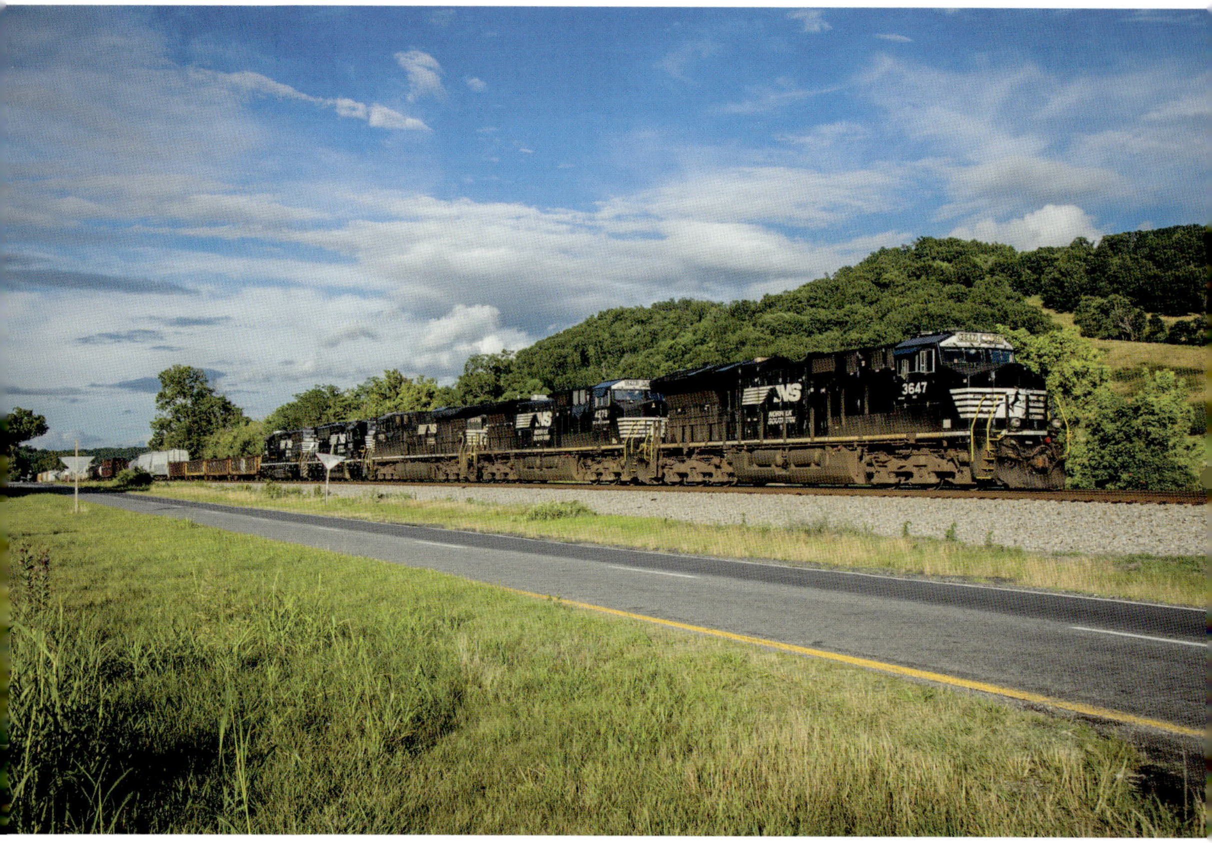
NS
3647

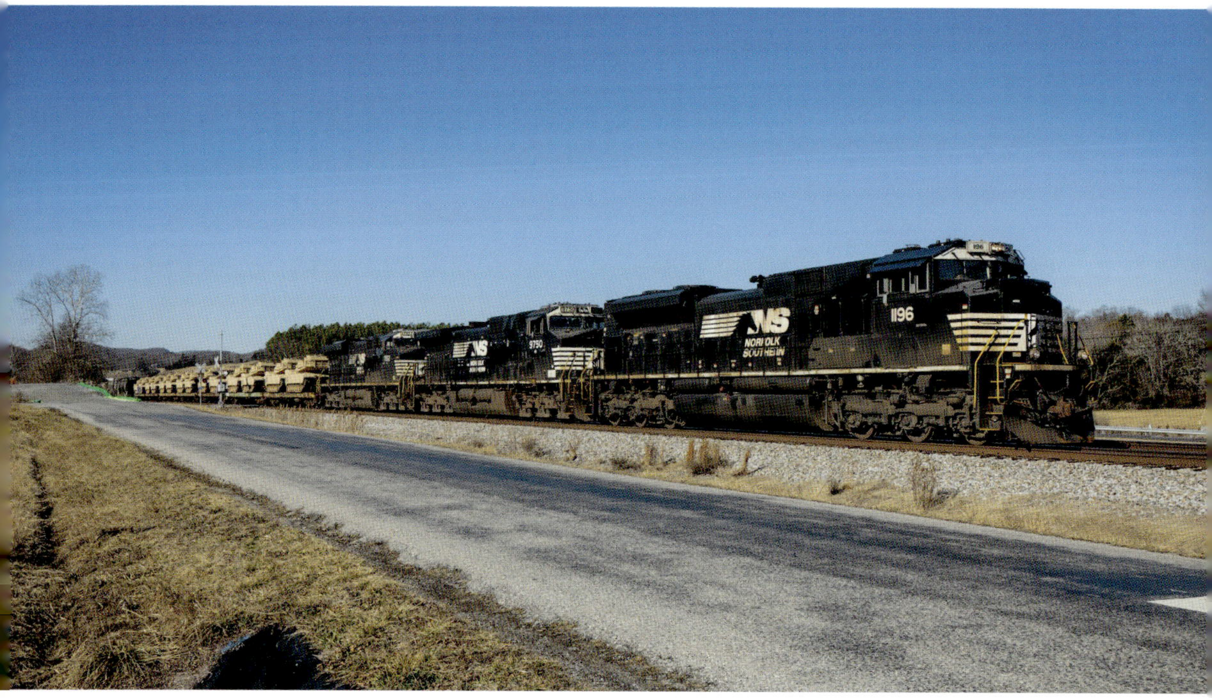
NS
NORFOLK SOUTHERN
1196

The day is drawing to a close and the sky is aglow from the setting sun as intermodal NS 234 heads east through Lafayette along the straightaway. The Dash 9, no. 9788, that leads the train was built in March 2003.

*Opposite above:* The first photographable location in Montgomery County is here in Lafayette, Virginia. This straight stretch of line parallels US Route 460 and passes by Rowe Furniture and the Elliston Vol. Fire Dept. An ET44AC leads train NS 19M west on a late July evening. Norfolk Southern ET44ACs, also known as Tier 4 GEVOs, are numbered 3600 to 3680 and were all built between May 2016 and February 2017.

*Opposite below:* An EMD SD70ACe and two GEs pull manifest NS 352 east between Cove Hollow Road and US 460 in Lafayette. Norfolk Southern special agents and the F.B.I. followed this train during its entire trip due to the transporting of military tanks just behind the power.

A pair of big EMDs pull hopper train NS 821 through a heavy snow in Lafayette, Virginia, as it heads toward the coal fields of West Virginia. This slight curve ends the straight stretch and the intermediate signal at Green Hill is just ahead of it.

*Opposite above:* An ES44AC is in charge of a long empty westbound coal hopper train as it tackles the straight stretch in Elliston. Unit 8110 was built in March 2012 for Norfolk Southern.

*Opposite below:* Vibrant fall colors are on display as *Blue Mane* AC44C6M 4005 lugs intermodal NS 23G east through Elliston, VA, at the start of the straightaway. 23G is a daily intermodal that runs from Louisville, KY, to Norfolk, VA.

The Greenhill intermediate signal can be seen in the distance as the Lehigh Valley heritage unit shoves the rear of a hopper train west through the Elliston/Lafayette area.

An overcast and dark February morning finds intermodal NS 217 rolling west through Elliston, Virginia, as it is about to duck under the US 460 bridge. A pair of ES44ACs lead the train.

Hay bales line the right of way in Elliston as hopper train NS 815 rushes west with the second AC44C6M rebuild on point, no. 4001, which was involved in a derailment with EMD SD70ACe 1036 where both units received extensive damage. Both were returned to service and 4001 was repainted in its original blue and gray paint scheme. This photo was taken not long after the unit was finished with repairs. No. 4001 is ex-Dash 9 no. 8879.

Passing what is known as the "big tree" by local railfans, NS 23G rolls east through Elliston, Virginia with an EMD SD70ACe leader.

Manifest NS 126 has just come down off of Christiansburg grade and now it is on more level ground as it crosses over the Blount Drive crossing in Elliston, Virginia. This unit, 3625, was involved in an incident where it hit a coal chute back in 2017. It was later repaired in Roanoke, Virginia, and the normal white "brow" around the number boards was not added back.

*Opposite above:* The leaves have all shed from the "big tree" in Elliston as eastbound manifest NS 392 rumbles by it with the Erie heritage SD70ACe in charge.

*Opposite below:* Here we see NS 41N again with a white face Dash 9 leading. This time, it is rolling west over the South Fork of the Roanoke River in Elliston. To the left and out of frame stands Big Spring Mill. The mill closed its doors in August 2022.

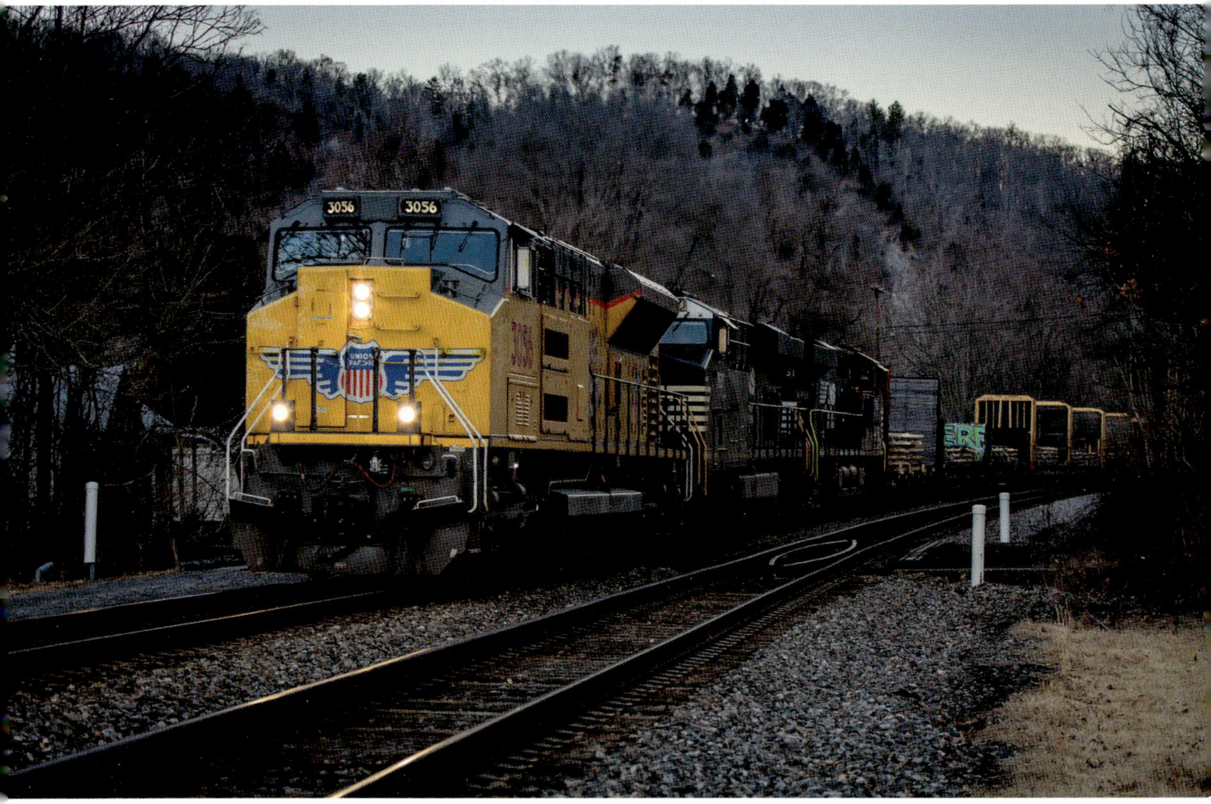
3056
3056
UNION PACIFIC

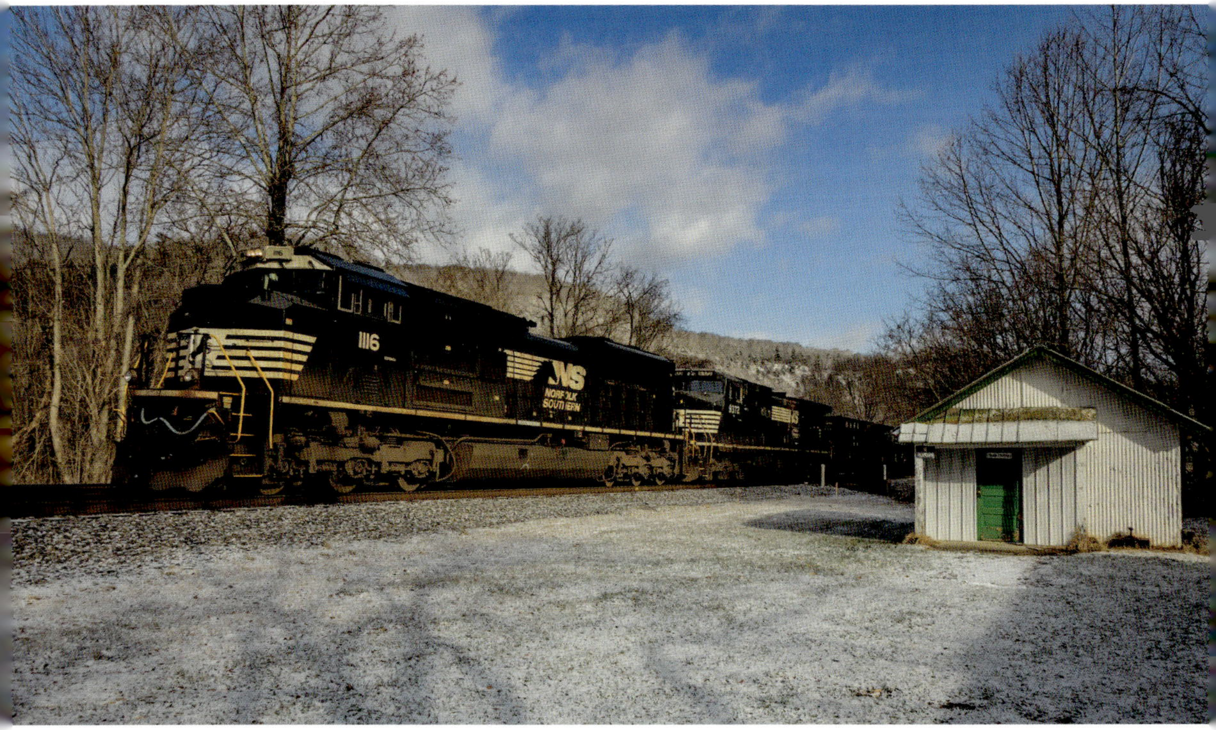
1116
NS
NORFOLK SOUTHERN

Seen stopped in Elliston is manifest/intermodal NS 202. The Lackawanna heritage unit is in the lead as the crew waits for dispatch to give the OK to start moving again. This is a popular place for Norfolk Southern dispatchers to stop eastbound trains until there is room in the yard as there are no crossings blocked.

*Opposite above:* Continuing westward through Elliston, we stop at Blount Drive. Seen from a local railfan's house, westbound hopper train NS 821 passes by. The building in the photo was built by the Norfolk and Western Railway and was moved from across the tracks many years ago. It is now owned by the railfan. This location in Elliston used to be a bustling place in the N&W days and included many signals with a three-track mainline and a passenger station that once stood near this spot. All that remains now of the past is this building, remnants of an old water tower, and rails still visible in the ground that went to a local industry. The depot was a freight and passenger depot in one and was demolished by Norfolk Southern.

*Opposite below:* A Union Pacific EMD Tier Four EMD ACe leads manifest NS 393 west through Elliston, Virginia, as the sun has set behind the mountain for the day.

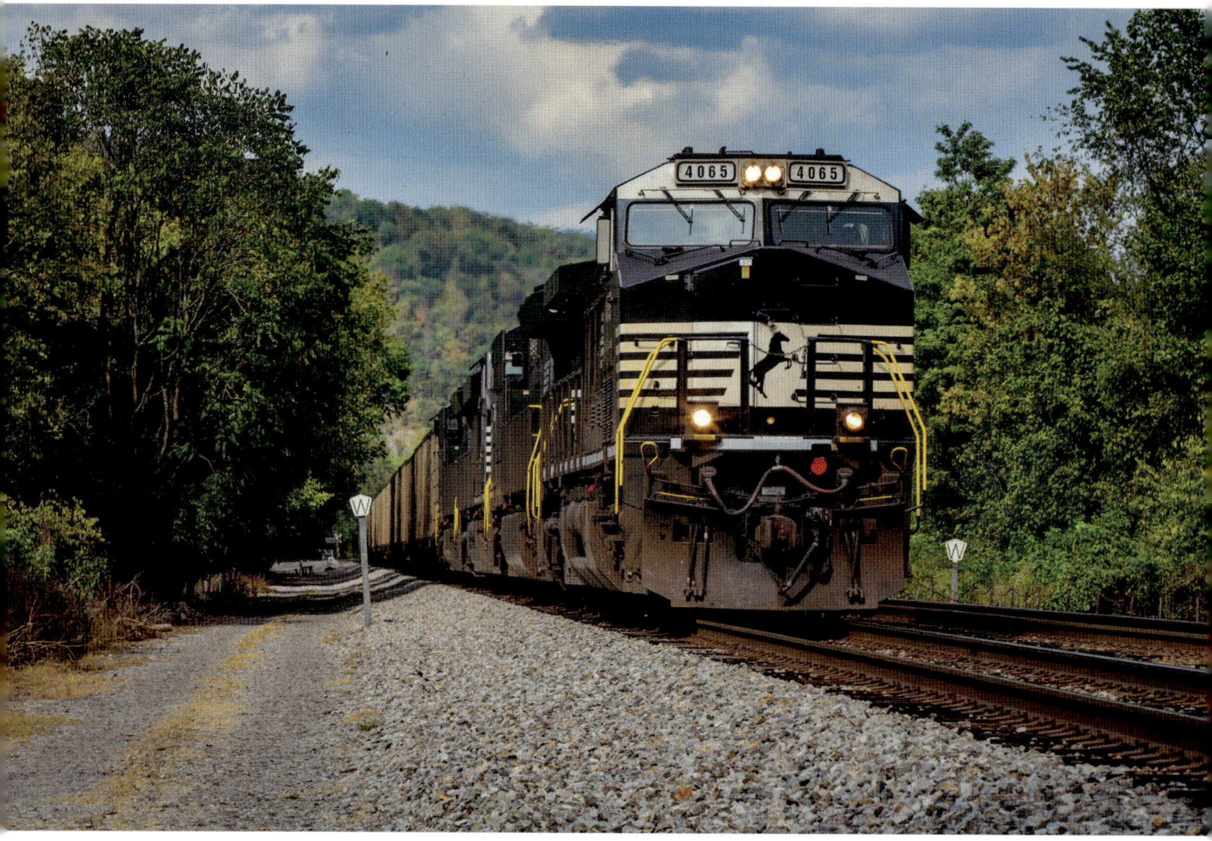
4065
4065

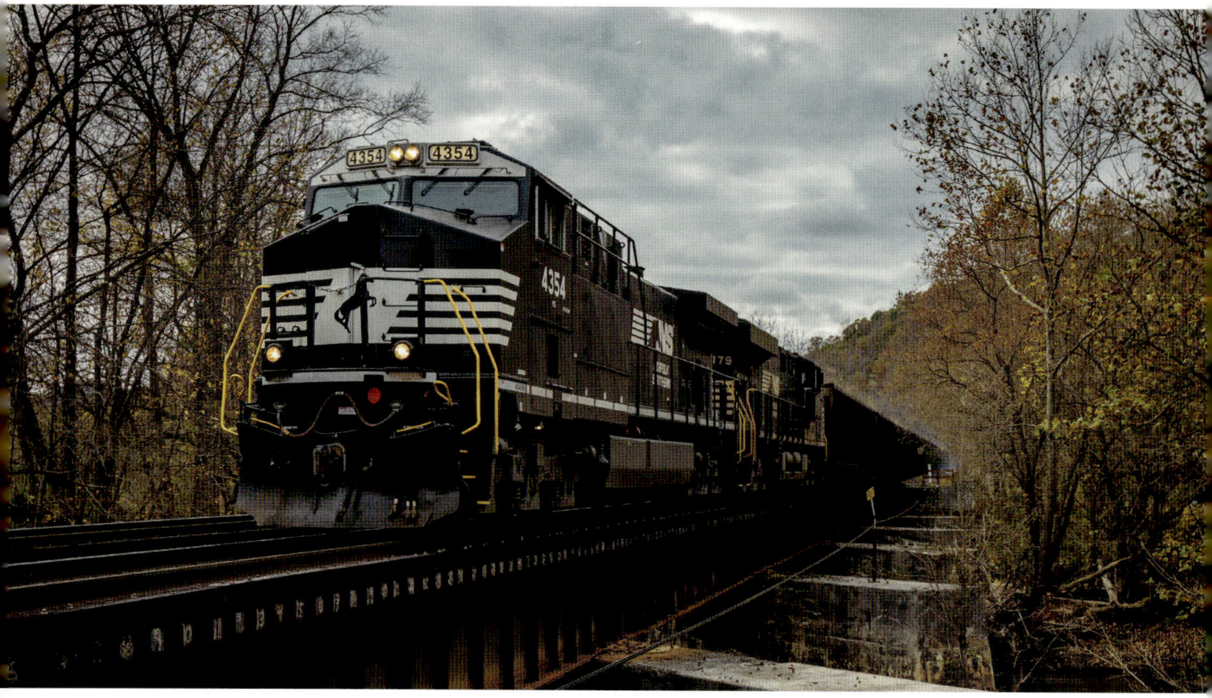
4354
4354
4354

Manifest NS 126 heads east by the Barnett Family Cemetery with the Virginian heritage SD70ACe out front. From this point west to Shawsville, the line is inaccessible as it is on the side of the mountain and in the trees. No. 126 is a daily manifest that operates from Chattanooga, Tennessee, to Roanoke, Virginia.

*Opposite above:* An AC44C6M leads an empty domestic coal hopper train west through Elliston, Virginia. It is seen splitting a pair of old Norfolk and Western whistle posts that have stood the test of time. Just ahead of the train is the start of the 1.5 percent Christiansburg grade where most trains slow down significantly while climbing toward Christiansburg, Virginia.

*Opposite below:* Loaded domestic coal train NS 756 rushes east across the south fork of the Roanoke River in Elliston. Brake smoke is flying, as seen in the background, as the train holds back as it has reached the bottom of Christiansburg grade. Evidence of a third main line can be seen to the right of the train.

This photo was taken from Blankenship Road, which is just behind Shawsville Middle School, and shows train NS 201 making its way up grade with a Dash 8 on point. This unit, no. 8449, was built for Conrail in November 1994 and was eventually acquired by Norfolk Southern. All of the GE Dash 8s are now off the roster. Around the curve in the background, there are the "Shawsville Cut" intermediate signals. At the location of the signal used to be a tunnel, before it was blasted away by NS.

*Opposite above:* From the opposite side of the small cemetery, NS 19M can be seen passing by with an ET44AC leader as it is about to tackle Christiansburg grade.

*Opposite below:* Next we come to Shawsville, Virginia, as the line continues up Christiansburg grade. Here at John Farm, the photo angles are almost endless in the small area, as evidenced by the amount of photos that will come after this one. Here intermodal NS 233 hurries west with an ES44AC in the lead. NS 233 is a daily intermodal that operates from Norfolk, Virginia, to Chicago, Illinois.

A Union Pacific SD70ACe is in charge of westbound intermodal NS 217 as it charges up Christiansburg grade and tackles the S-curve at John Farm. This train had units from Union Pacific, Norfolk Southern, and BNSF.

The Savannah and Atlanta heritage SD70ACe pulls a westbound coal hopper train through the S curve at John Farm just before midnight on a cold March night.

Seen from the field on the opposite side of the tracks, late day sun shines down on the Reading Lines heritage SD70ACe as it leads eastbound manifest NS 126. The Reading Lines Railroad ran through Southeast Pennsylvania and neighboring states. The railroad was operated from 1924 to 1976 when the line was absorbed by Conrail, which was eventually absorbed by Norfolk Southern.

A warm spring night finds manifest NS 18M rushing down grade through Shawsville with a trio of GE products on point. Every spring, the field at John Farm is covered with yellow rapeseeds.

We move just a little west, but still at John Farm for this shot and the next couple shots. Seen from a higher viewpoint, NS 12Z glides east down the mountain with a brand new AC44C6M rebuild on point. Summer is a busy time here at John Farm for the local landowners.

*Opposite above:* John Farm offers so many different angles for photos, but this one is the most desired. Seen from the fence surrounding an old cemetery on the hill, westbound intermodal NS 201 rushes upgrade with a pair of GE products making easy work of the train. This fence is quite old, and sections have fallen down over the years unfortunately.

*Opposite below:* Seen from inside the cemetery at John Farm, eastbound intermodal NS 234 rushes by just after the sun has set over the mountains for the day. A trio of GE products are in charge of the train with an AC44C6M in the lead.

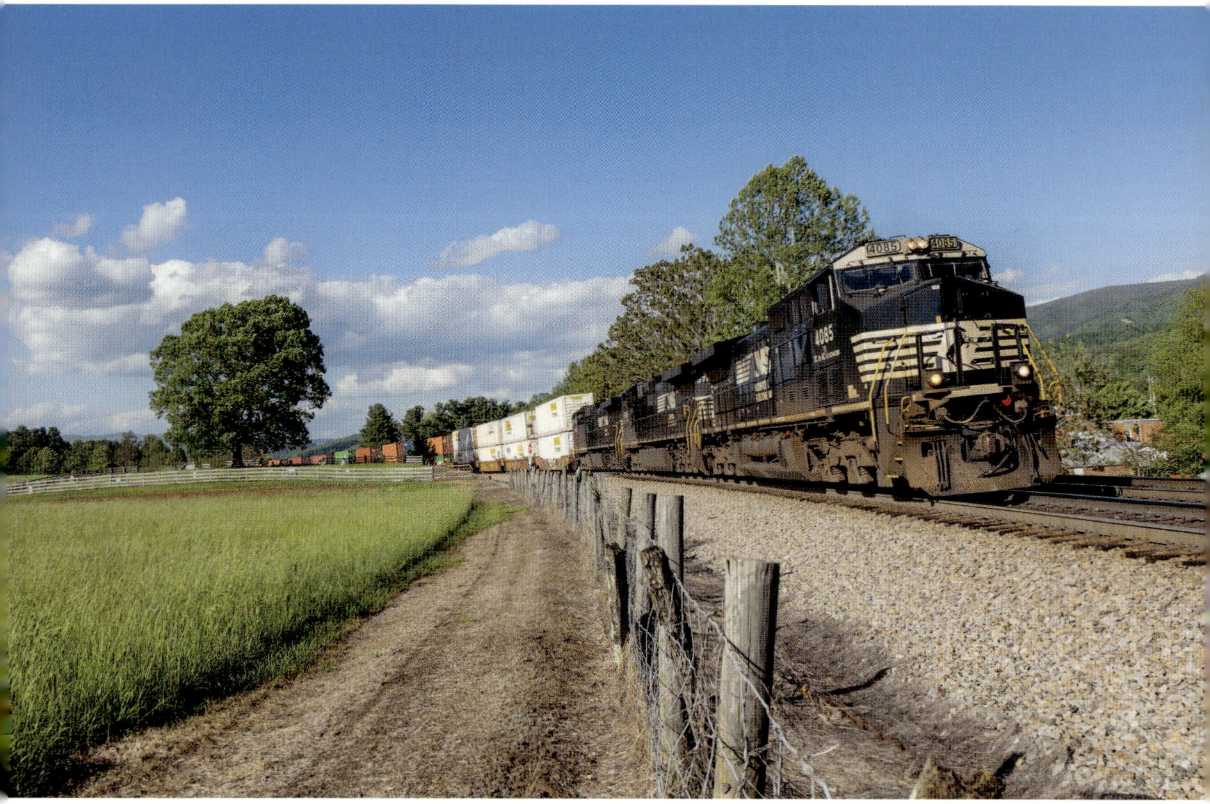

W

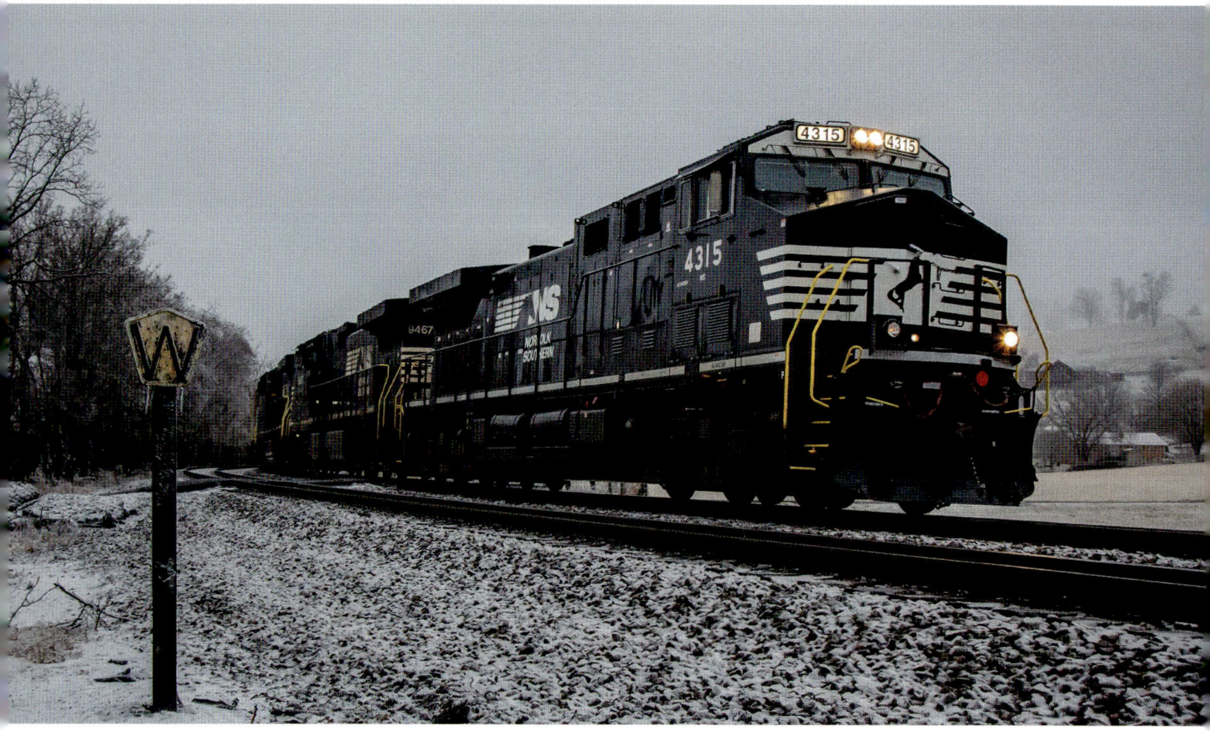

The smell of wood smoke fills the air on a cold December night as manifest NS 352 rushes east by John Farm in Shawsville with the Lehigh Valley heritage unit on point.

*Opposite above:* An AC44C6M rebuild leads westbound intermodal NS 29G along the fence line at John Farm in Shawsville. The tree to the left of the train is in the center of the cemetery where two previous vantage points were from.

*Opposite below:* A brand-new GE AC44C6M rebuild leads intermodal NS 202 east past an old Norfolk and Western whistle post at John Farm on a cold winter evening. Unit 4315 was originally NS Dash 9 no. 9158 before being rebuilt.

AC44C6M 4373 leads an empty domestic coal hopper train west as it approaches the Newtown Road crossing in Shawsville just past John Farm. Shawsville began life as a railroad stop in the 1850s for the nearby Alleghany Springs resort. The town got its name from Virginia's chief engineer at the time, Charles B. Shaw.

After rounding the curve and leaving John Farm behind, a westbound empty grain train with a Dash 9 leader approaches the Newtown Road crossing. Newtown is a small community located in Shawsville, Virginia.

Eastbound manifest NS 18M is about to glide across the Newtown Rd crossing with a Union Pacific GE ES44AC in charge on a late July evening. To the left used to stand an original Virginia and Tennessee passenger station. The station stopped seeing passengers in the 1950s and was rented out by a farm supply store. The station then burnt down on a Halloween night in the 1970s. The concrete milepost for N281 can be seen to the right of the train.

The gray and white Monongaleha heritage unit leads an empty coal hopper train west by the Newtown community sign in Shawsville, Virginia. The Monongahela Railway operated in Pennsylvania and West Virginia from 1900 to May 1993 when it was merged into Conrail. The railway primarily transported coal.

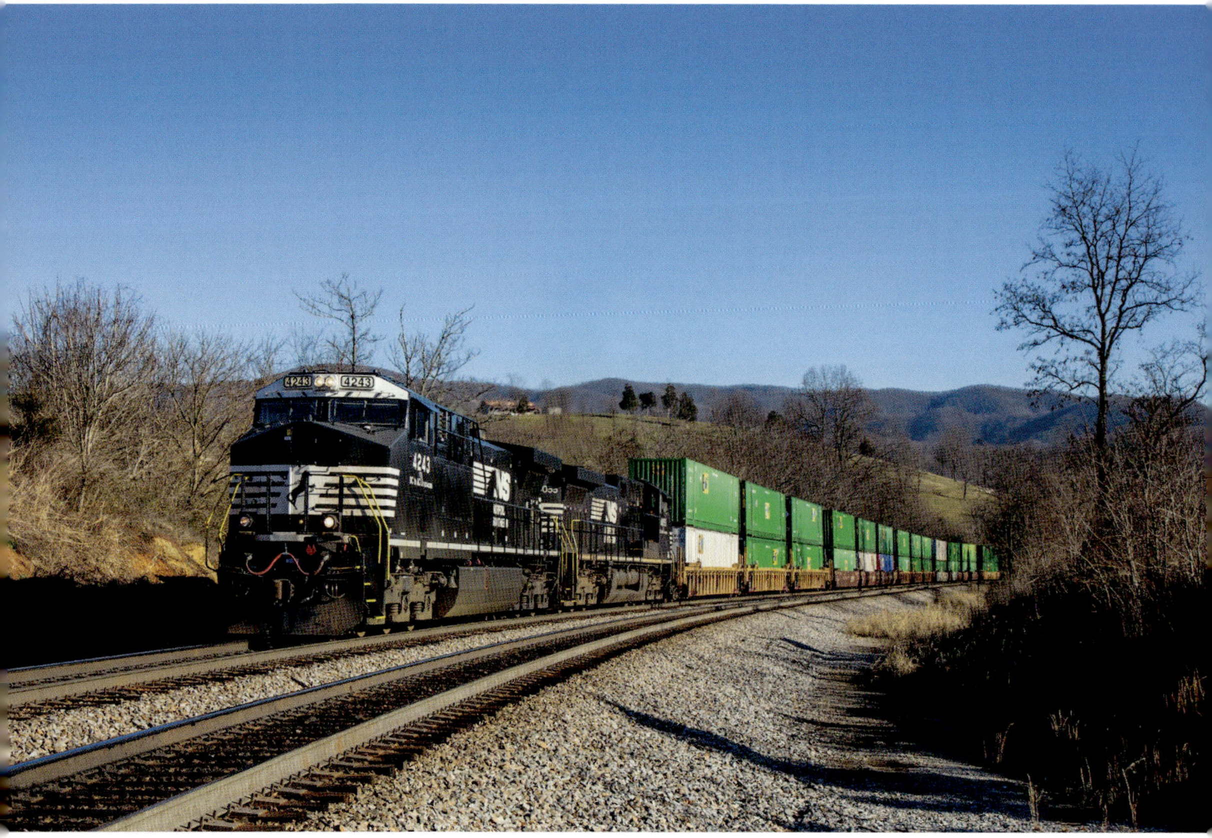
4243
4243
4243

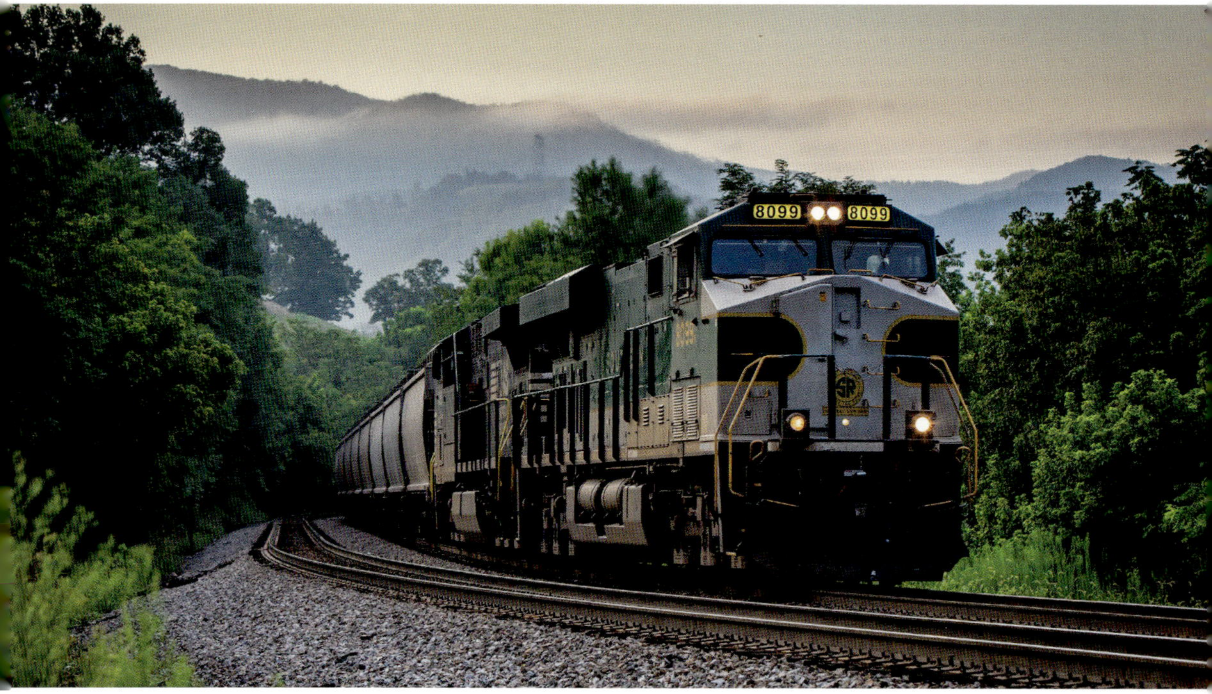
8099
8099

An SD70ACU lugs manifest NS 19M up Christiansburg grade in Shawsville at the Basham Hollow Road crossing.

*Opposite above:* More picturesque views can be found west of Newtown Road here at the Basham Hollow Road crossing in Shawsville. Intermodal NS 201 is about to rumble over the crossing with a pair of AC44C6M rebuilds on point on a sunny winter afternoon.

*Opposite below:* On an early August morning, before the sun has come up over the mountain, empty grain train NS 51X rumbles west with the Southern heritage unit. The Southern Railway was a Class 1 railroad that operated from 1894 to 1982. It merged with the Norfolk and Western in 1982 to form the modern-day Norfolk Southern.

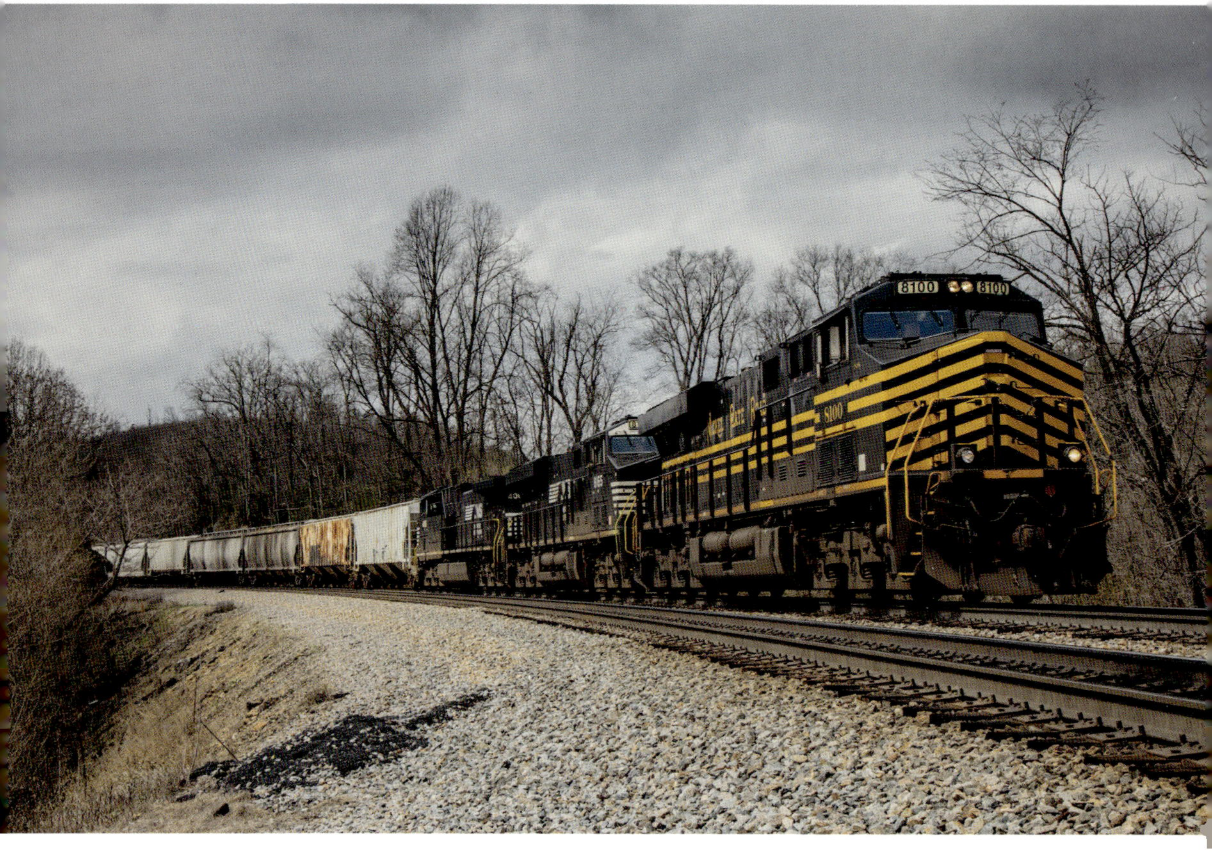
8100
8100

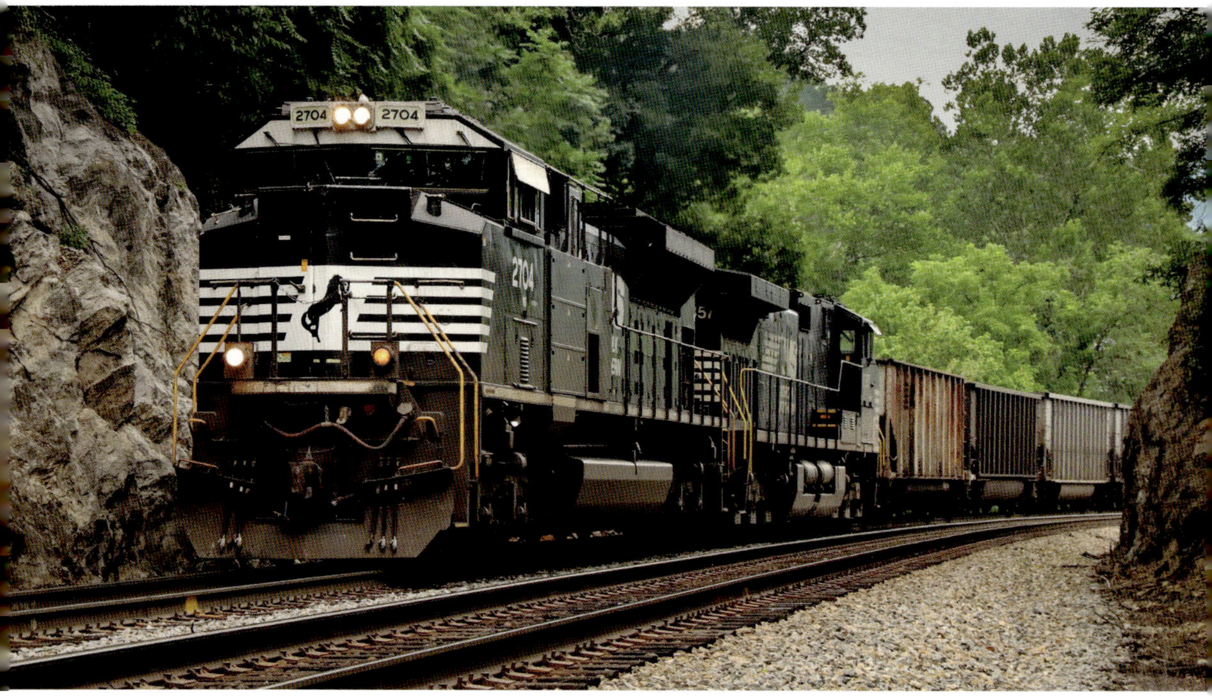
2704
2704
2704

The last bit of day light shines down on eastbound manifest NS 392 as it rushes toward Ryan Road in Shawsville.

*Opposite below:* Eastbound manifest NS 18M rushes down grade as it approaches the Basham Hollow Road crossing in Shawsville, Virginia, with the Nickel Plate Road heritage unit on point.

*Opposite below:* The next location we come to is the Ryan Road crossing in Shawsville. The rocks were blasted away to make way for the line in this location. Hopper train NS 811 heads west here with an EMD SD70M-2 in charge. This unit was built in 2006 and has since been retired along with every other SD70M-2 NS owned. This was the last SD70M-2 I would photograph in service.

Four Norfolk Southern units drag empty coal hoppers west through control point Arthur in Shawsville. This control point was named after the individual who sold the land to the Norfolk and Western Railway. Control point Arthur is milepost 282.2.

Framed by a budding Redbud tree, intermodal NS 22A rumbles west through control point Arthur. The gravel Sparrow Road can be seen in the foreground.

An AC44C6M rebuild and an old logo Dash 9 bring manifest NS 392 east through the signals at control point Arthur after a heavy snow.

With fresh snow on the ground, eastbound intermodal NS 202 rolls toward control point Arthur in Shawsville. An AC44C6M leads the train.

A freshly completed AC44C6M rebuild leads empty grain train NS 51X through the signals at control point Arthur as the redbuds are blooming. No. 4333 was rebuilt from NS Dash 9 no. 9010 in February 2020.

AC44C6M 4004 leads manifest 19M west just past Arthur in Shawsville. This crossing is a private road crossing off of US-460.

Intermodal NS 202 drifts down the grade through Shawsville in a light rain as it heads toward Arthur with an ET44AC in the lead.

A big ex-Conrail EMD SD80MAC lugs a hopper train up the mountain to our next location at the Friendship Road crossing in Shawsville. All of the 80MACs have been retired and most, if not all, are being scrapped.

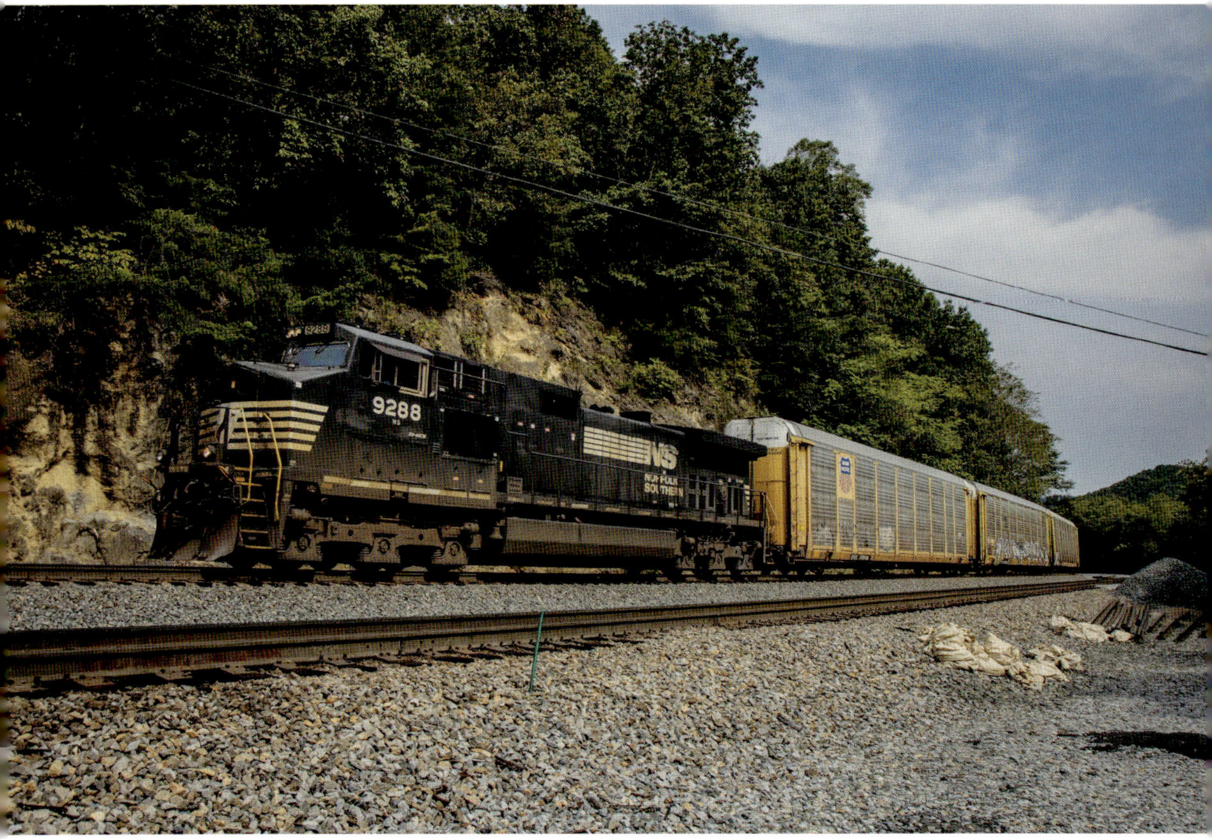
9288
NS
NORFOLK
SOUTHERN

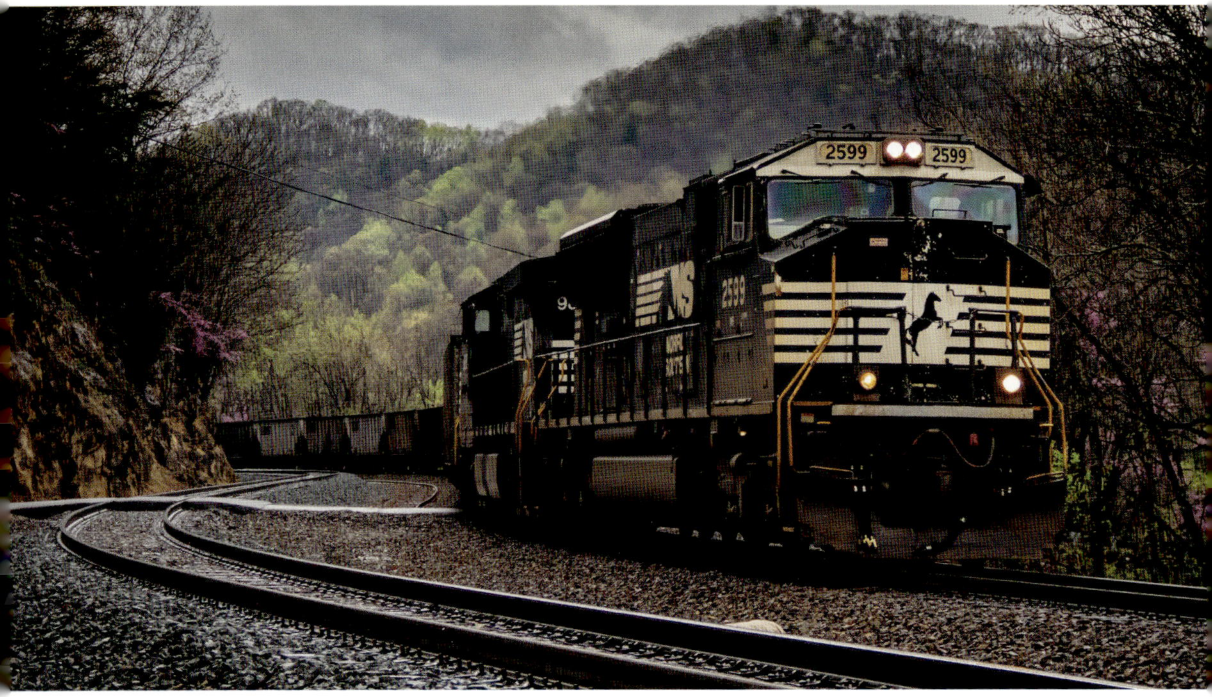
2599
2599
NS
2599

A CSX AC44CW still in old YN2 paint leads eastbound intermodal NS 234 as it cruises down the 1.5 percent grade toward the Friendship Road crossing.

*Opposite above:* A wider view from the Friendship Road crossing shows an autorack train, NS 29J, rolling west up the grade with a single old logo Dash 9 leading the way. No. 9288 has now been rebuilt as AC44C6M no. 4426.

*Opposite below:* Seen from the other side of the crossing at Friendship Road, an EMD SD70M is in the lead of hopper train NS 811 as it rumbles west up Christiansburg grade. No. 2599 was built in April 2003 and was retired in early 2020 along with all other SD70Ms owned by Norfolk Southern.

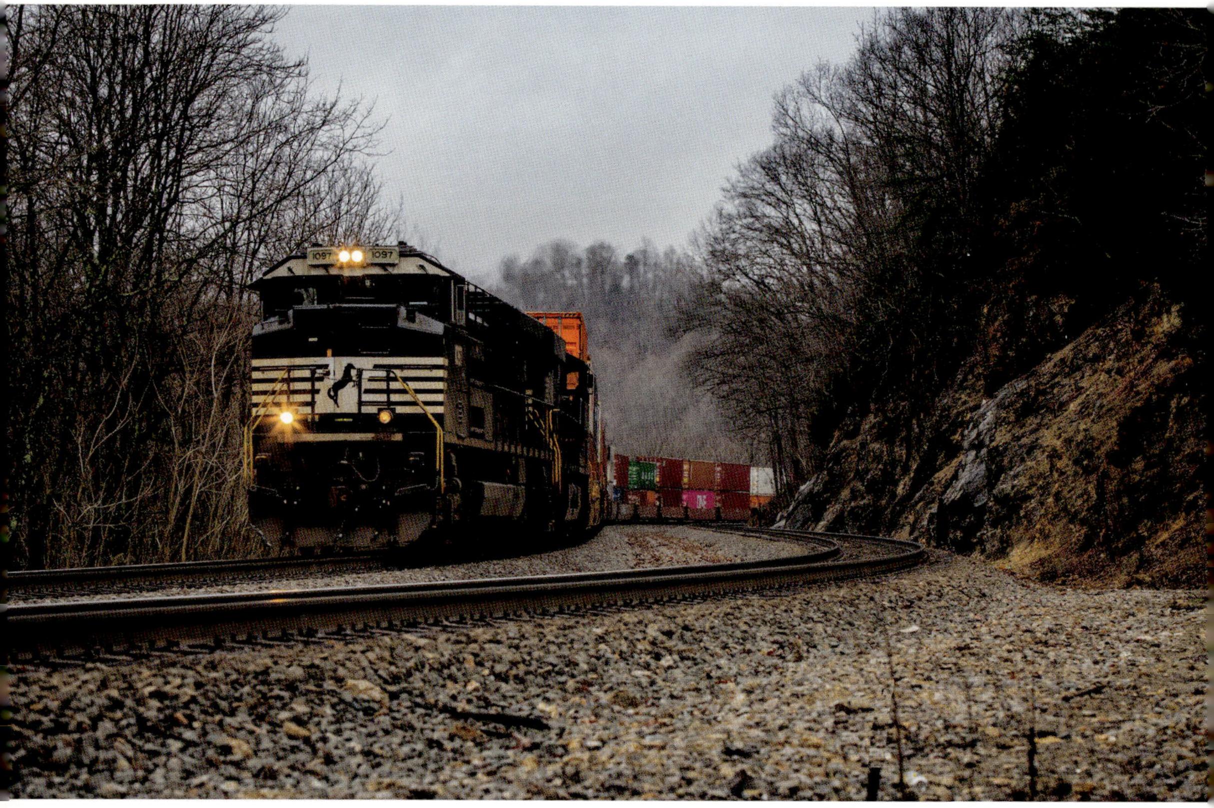
1097
1097

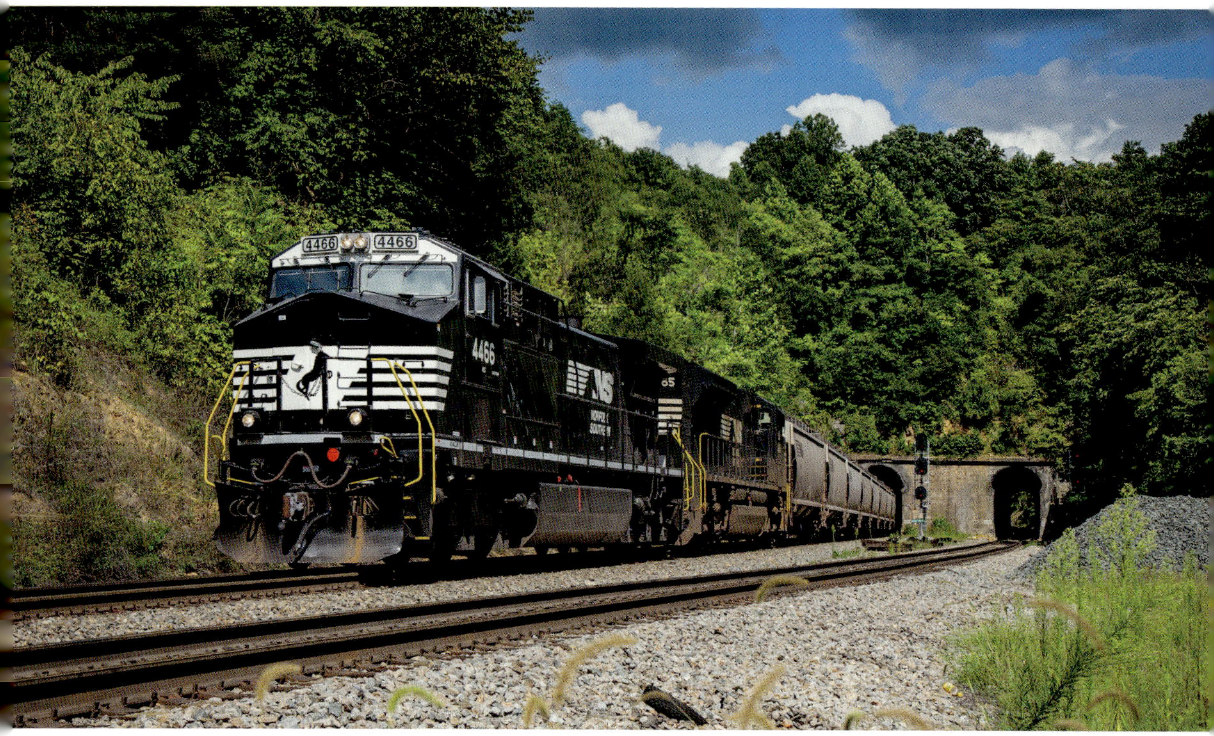
4466
4466
4466

The Norfolk and Western heritage unit is on "home rails" as it leads a long train of empty hoppers out of Montgomery Tunnel and toward the coal fields of West Virginia. The N&W operated from 1870 to 1982 when it merged with the Southern Railway to form the current day Norfolk Southern. The N&W was famous for building many of its own steam locomotives including the famous J-class 4-8-4s and A-class 2-6-6-4s.

*Opposite above:* The final photo near the Friendship Road crossing shows intermodal NS 23G coasting down Christiansburg grade with an SD70ACe showing the way.

*Opposite below:* Empty grain hopper train NS 55W pops out of Montgomery tunnel at control point Montgomery (the crossover is located at the opposite side of the tunnel) between Shawsville and Christiansburg, Virginia, off of Den Hill Road. The well-known nighttime rail photographer O. Winston Link made this location famous with his photos of steam giants rolling out of the tunnel. Double stack intermodals have only recently been able to go through both portals of the tunnel with the ceiling being raised during the Heartland Corridor Project. The project improved rail operations between Norfolk, Virginia, and Chicago, Illinois.

Brand new AC44C6M no. 4560, ex-Dash 9 no. 9320, leads eastbound intermodal NS 23G downgrade as it approaches control point Montgomery. Control point Montgomery is milepost 284.6.

Seen just around the curve from Montgomery tunnel, intermodal NS 217 heads west with a Dash 9 on point.

An empty grain train heads west up the mountain just past control point Montgomery. A Burlington Northern Santa Fe unit pulls the train under Interstate 81.

Manifest NS 126 rolls down the grade toward Montgomery tunnel. The Jersey Central Lines heritage unit rides second behind another SD70ACe.

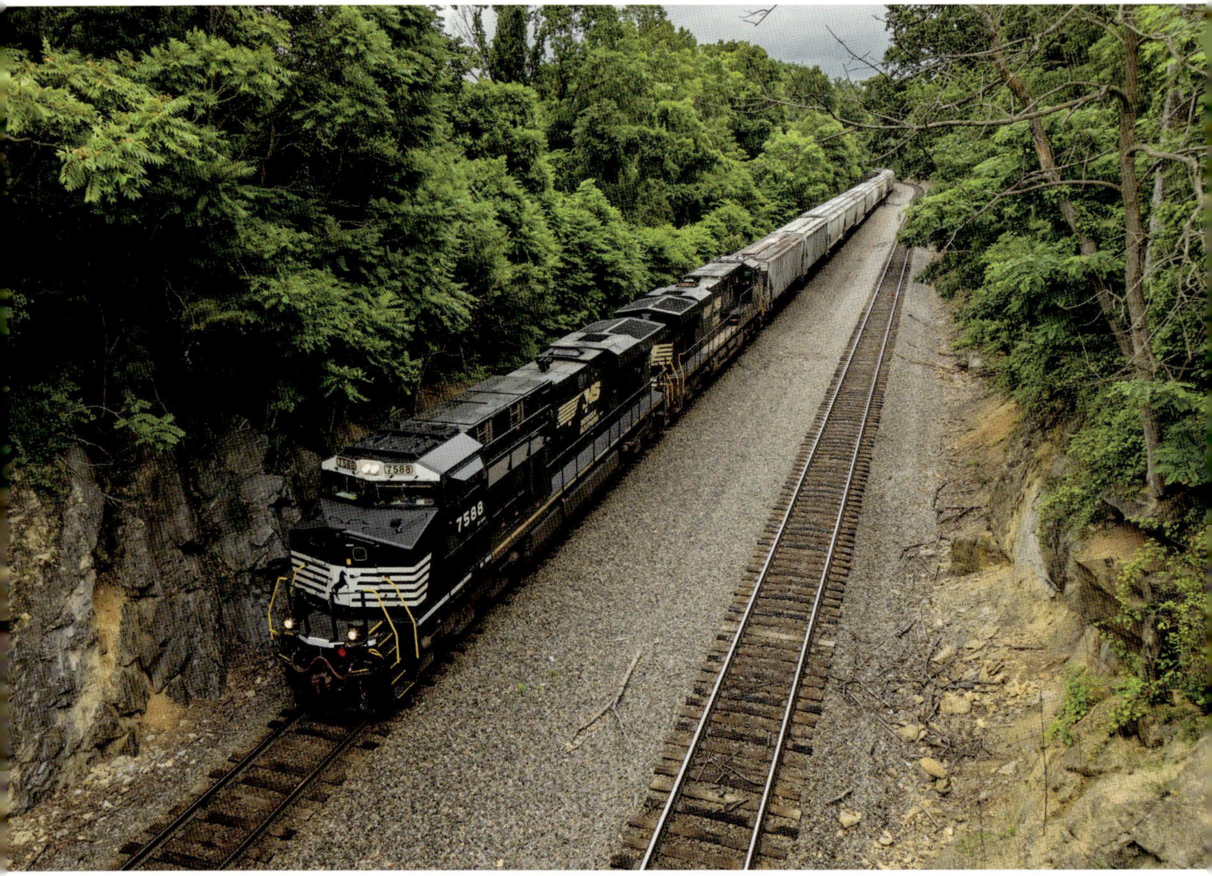
7588
7588
7588

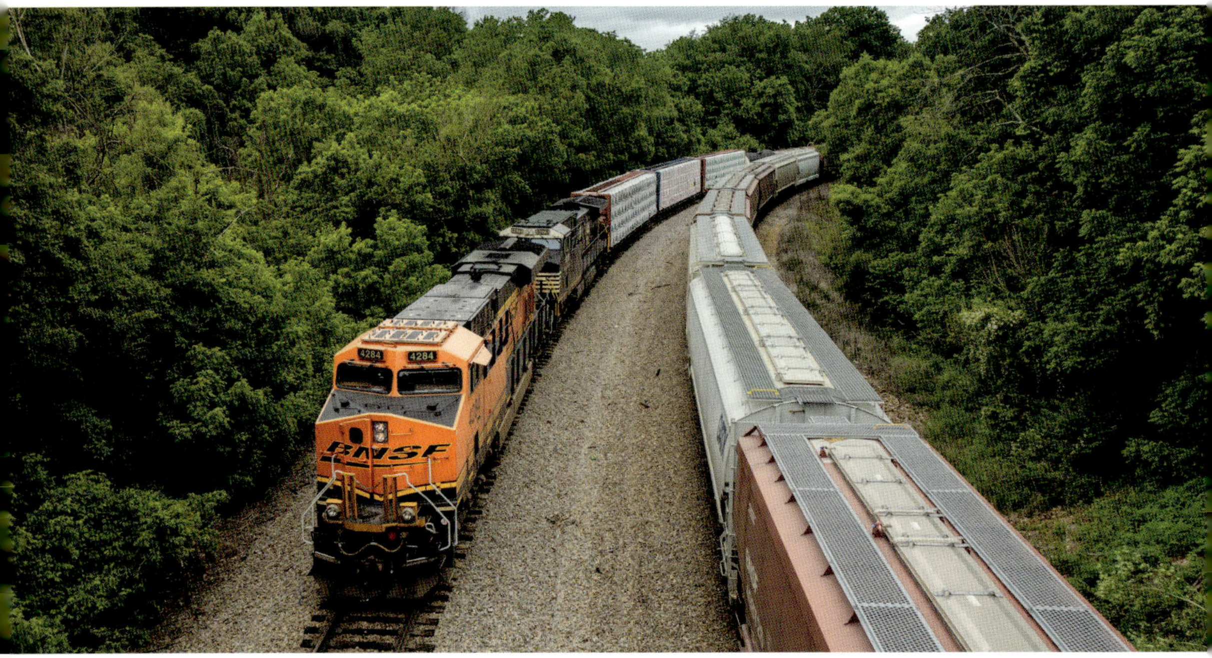
4284
4284
BNSF

We have arrived at the crest of the grade at the district's namesake of Christiansburg. The Pennsylvania heritage unit leads train NS 393 west by a depot in town. This depot was built in 1906 as a passenger depot and is now used by NS maintenance. This is where westbound trains will begin their descent down Charleston grade.

*Opposite above:* An empty grain train, NS 55W, fights up the grade through Christiansburg not far from the crest of the grade. A pair of ES44DCs lead the train as it is about to duck under US 460 bypass.

*Opposite below:* A BNSF unit leads eastbound manifest NS 352 down Christiansburg grade between Christiansburg and Montgomery. Here it is about to dive under US-460 as it passes its westbound counterpart NS 353. 352 is a daily manifest that operates from Portsmouth, Ohio, to Linwood, North Carolina.

Manifest NS 353 is seen passing westbound by the Cambria depot in Christiansburg. This view is from the third floor of an antique shop and shows Depot Street.

*Opposite above:* A side view of the Lackawanna heritage unit shows just how worn these paint schemes have become since their unveiling back in 2012. The unit is shown leading westbound manifest NS 15T through the eastbound signal at control point Christiansburg. 15T has since been abolished through this area. Control point Christiansburg is milepost N289.7.

*Opposite below:* At the crest of Christiansburg grade lies the Cambria depot. The original depot that was in this location was built in 1856 for the Virginia and Tennessee Railroad. It was burned down by Union forces in 1865. A new depot, the current day depot, was built in 1865 and now currently houses a used book and toy store.

A pair of Canadian National GE units are the power for an empty ethanol train as it crests Christiansburg grade and passes the Cambria depot just after nightfall.

Just before midnight on a warm October night, manifest NS 126 heads east through Cambria. The Original Norfolk Southern heritage unit leads the long freight train. The original Norfolk Southern Railroad ran from Norfolk, Virginia to Elizabeth City, North Carolina and was founded on January 1, 1883.

On an overcast Saturday morning in July, westbound manifest NS 127 thunders under the signal at control point Pelton in Christiansburg. A Union Pacific EMD SD70M is in rear DPU (distributive power unit) duty as it shoves on the rear of the heavy manifest. No. 127 is a daily manifest that operates from Roanoke, Virginia, to Chattanooga, Tennessee.

Eastbound stack and autorack train NS 23G has just crested Charleston grade and will now descend Christiansburg grade. It has the Cambria depot in its windshield and Christiansburg grade ahead of it. A mural was recently painted at the sidewalk over Crab Creek depicting the N&W J Class 611 and the Cambria depot.

Rolling west on main one just past control point Pelton is intermodal NS 22A. The New York Central heritage unit is in charge of the intermodal as it is about to dive under North Franklin Street in Christiansburg.

*Opposite above:* An old logo Dash 9 leads a combined grainer and manifest westbound under the signal at control point Pelton in Christiansburg. This is the start of a westbound train's descent down Charleston grade.

*Opposite below:* Export coal hopper train NS 821 sweeps the curve just past control point Pelton in Christiansburg during a measurable snow storm. An EMD SD70ACU leads the long empty train west toward Bluefield, West Virginia.

Eastbound intermodal NS 23G swings a curve between Christiansburg and Vicker, VA, with the green Illinois Terminal heritage unit in the lead. The Illinois Terminal operated from 1896 to 1982 when it was absorbed by the Norfolk and Western Railway. The railroad operated around 550 miles of rail between St. Louis, Missouri, to Danville, Illinois.

*Opposite above:* Manifest NS 393 passes west by Railroad Avenue, which is just west of control point Pelton. Railroad Avenue is a dead-end street that ends at the tracks not far in front of the train. A pair of GE products and an EMD lead the train. The road bridge in the background was the vantage point for the last three photos.

*Opposite below:* A Canadian National ES44AC is in charge of intermodal NS 22A as it glides down Charleston grade on main two alongside Chrisman Mill Road in Christiansburg.

A pair of ES44ACs are seen cruising west down Charleston grade with NS 69R, empty rock train, in Vicker, Virginia, on a rainy fall morning. Vicker was named in honor of pioneer settlers, the Vicars.

Another iconic location on the eastern end of the Christiansburg district is the next location of Vicker, Virginia, just west of the last photo. A Norfolk and Western coaling tower still stands tall as hotshot intermodal NS 217 rolls west under it with the Pennsylvania heritage unit. It is also about to pass a vintage N&W whistle post.

An ET44AC is in the lead of intermodal NS 23G as it thunders east under the old coaling tower in Vicker. Here it is seen tackling Charleston grade on a sunny September morning. This was a bustling place back in Norfolk and Western steam days with steam locomotives receiving more coal and topping off on water here.

The first AC44C6M rebuild, 4000, leads manifest NS 16T east underneath the N&W era coaling tower in Vicker. 16T has since been abolished here and was replaced with NS 126.

We now come to an important part of the line. Ahead of hopper train NS 757 lies the wye at Walton. Off of main two, the line continues west toward Bluefield, and splitting off of main one, the line will go west onto the Pulaski district toward Bristol, Tennessee. No. 757 will continue west along the Christiansburg district.

Looking west toward the wye at Walton, we see eastbound intermodal NS 234 continuing toward Norfolk, Virginia. To the left, you can see where the line splits and forms the Pulaski District. Control point Walton is milepost 297.6.

Moving along now to Pepper, Virginia. Not far past the wye at Walton, the line is single tracked here until it reaches Narrows, Virginia. This is a great location to catch an eastbound train in the sun almost any time of the day. Dash 9 no. 9551 leads train 234 east through Pepper, Virginia, as it is about to roll underneath Peppers Ferry Road. No. 9551 was built for Norfolk Southern by General Electric in December 2000.

Now, moving into Giles County, Virginia, we see the red maned AC44C6M no. 4002 as it is serving as a rear distributive power unit shoving westbound hopper train NS 763 toward Eggleston, Virginia. Here it is running alongside the gravel Eggleston River Road, which lies between the mainline and the New River. The red stripes, or "mane," signify that this AC44C6M was a rebuild completed in the shops of Roanoke, Virginia.

Hopper train NS 815 works through the S-curve in Pembroke, Virginia, on an early spring afternoon. This location is only a couple hundred yards west of Pembroke tunnel off of River Road.

*Opposite above:* A view from the Eggleston Road bridge shows intermodal NS 217 storming through the mountains of South Western Virginia. The Penn Central heritage unit shows the way as it passes a house with a great view of the old Norfolk and Western mainline.

*Opposite below:* Pembroke tunnel can be seen from above on River Road in Pembroke, Virginia, as we continue past Eggleston. On this day, a single SD70ACe leads manifest 353 west out of the east portal of the tunnel.

The sun has not quite risen over the mountains of Southwestern Virginia as intermodal NS 234 rushes east through Ripplemead, Virginia, before it dives under US 460. Ripplemead is a small community that lies along the New River.

*Opposite above:* Loaded domestic coal train NS 744 ducks under the signal in Pembroke, VA. Control point Pembroke is milepost 320.6. No. 4433 was rebuilt from Dash 9 no. 9169 in March 2021.

*Opposite below:* Just west of Pembroke is Ripplemead, Virginia. This view comes from a bridge on US 460 and shows hotshot intermodal NS 217 heading west around a curve at mile post N323 with the Jersey Central Lines heritage unit on point. This is milepost 323.0.

Westbound manifest NS 393 heads past the intermediate signals in Pearisburg, Virginia, with an AC44C6M rebuild on point. Pearisburg was founded in 1808 and was named after landowner George Pearis, who donated 50 acres of his land to be used for the town.

Exhaust fills the air as three GE units throttle up after a brief stop in Pearisburg, Virginia, waiting for a signal to continue east with manifest NS 18M. To the right is a public boat ramp to the New River.

A pair of Union Pacific GEs pull an empty ethanol train west through Narrows, Virginia. Across this small crossing is a red barn and an AM radio service. Narrows was named for the narrowing of the New River that runs through the town and was settled around 1778.

The sun pops out of the clouds at just the right time as a nice-looking pair of ES44ACs bring intermodal NS 217 west through Narrows, Virginia, alongside Narrows Road. The conductor gives a wave as they head toward Williamson, West Virginia, for their next crew change. No. 217 terminates in Chicago, Illinois.

The first SD70ACC rebuild, no. 1800, leads hotshot intermodal NS 217 west through Narrows, Virginia, alongside Narrows Road. These rebuilds were performed by Progress Rail, which is a subsidiary of CAT. This is the reason for the yellow on this unit and 1801. SD70ACCs are former standard cab SD70s.

Here we see export coal train NS 822 rocketing east at the start of the Whitethorne district in Narrows. The district begins just around the curve to the left and branches off of the Christiansburg district, which is seen in the foreground. Most eastbound coal, grain, manifests, and ethanol take the Whitethorne due to the easier grades toward Roanoke. Double stack intermodals cannot take the Whitethorne due to lower tunnel clearance. The signals at control point Narrows (milepost N332.8) are visible in the background.

The Penn Central heritage unit is in the lead of NS 217 as it rushes over top of MacArthur Lane in Narrows. Out of frame to the left is an old depot that now houses Norfolk Southern maintenance equipment. Straight ahead under the bridge is downtown Narrows.

The Norfolk and Western Railway logo is still showing proudly on the bridge over MacArthur Lane.

A loaded grain train is seen passing the slide fence in Narrows, Virginia, with an ES44AC leader as it heads east toward Roanoke for its next crew change. This train will cross over to the Whitethorne district just ahead after it passes over MacArthur Lane.

The Stock Pen Mountain Road crossing in Narrows, Virginia, is the next location as we continue traveling west. Here we see a pair of Norfolk Southern units shoving the rear of an eastbound coal drag through the mountains.

Loaded coal train NS 820 rushes east through the mountains of Southwestern Virginia in Narrows, Virginia. The train is continuing its trek toward Norfolk, Virginia, for the coal to be exported. An ES44DC leads the heavy train.

Eastbound manifest NS 392 finds a small pocket of late evening sunlight just near Stock Pen Mountain Road in Narrows, VA.

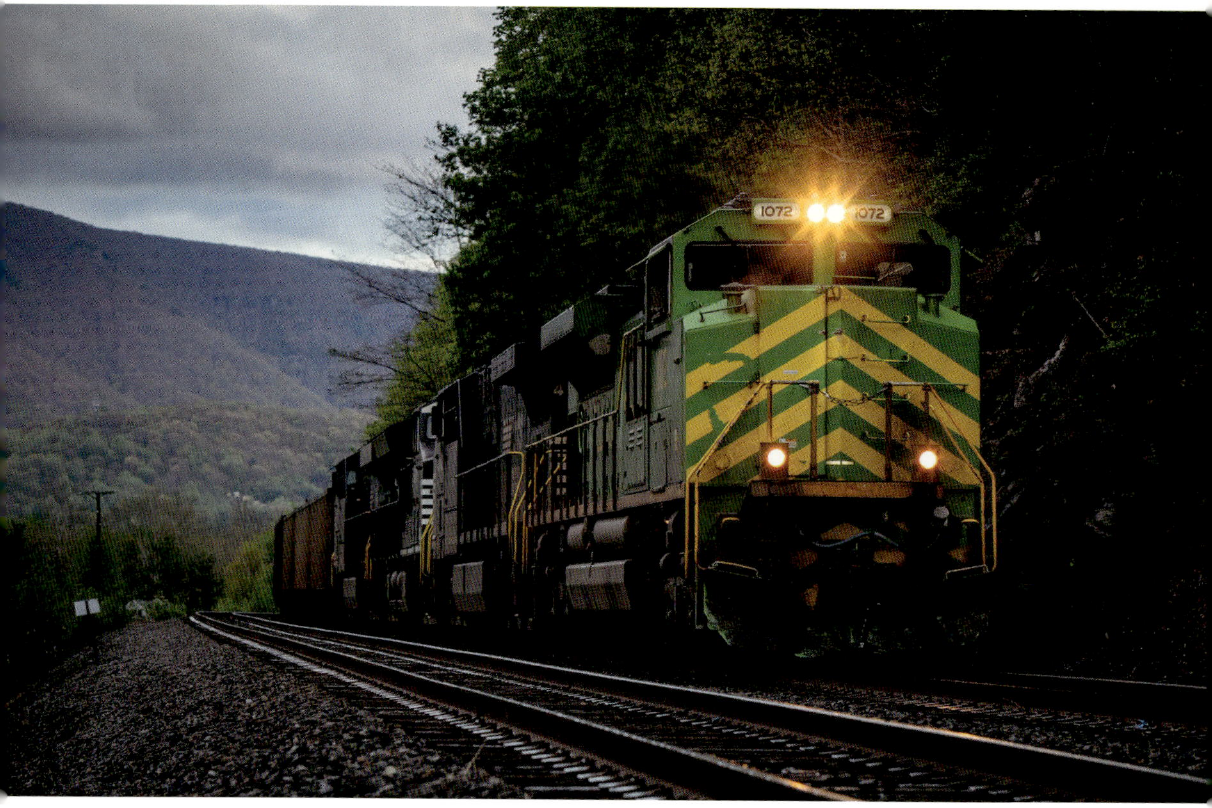
1072
1072

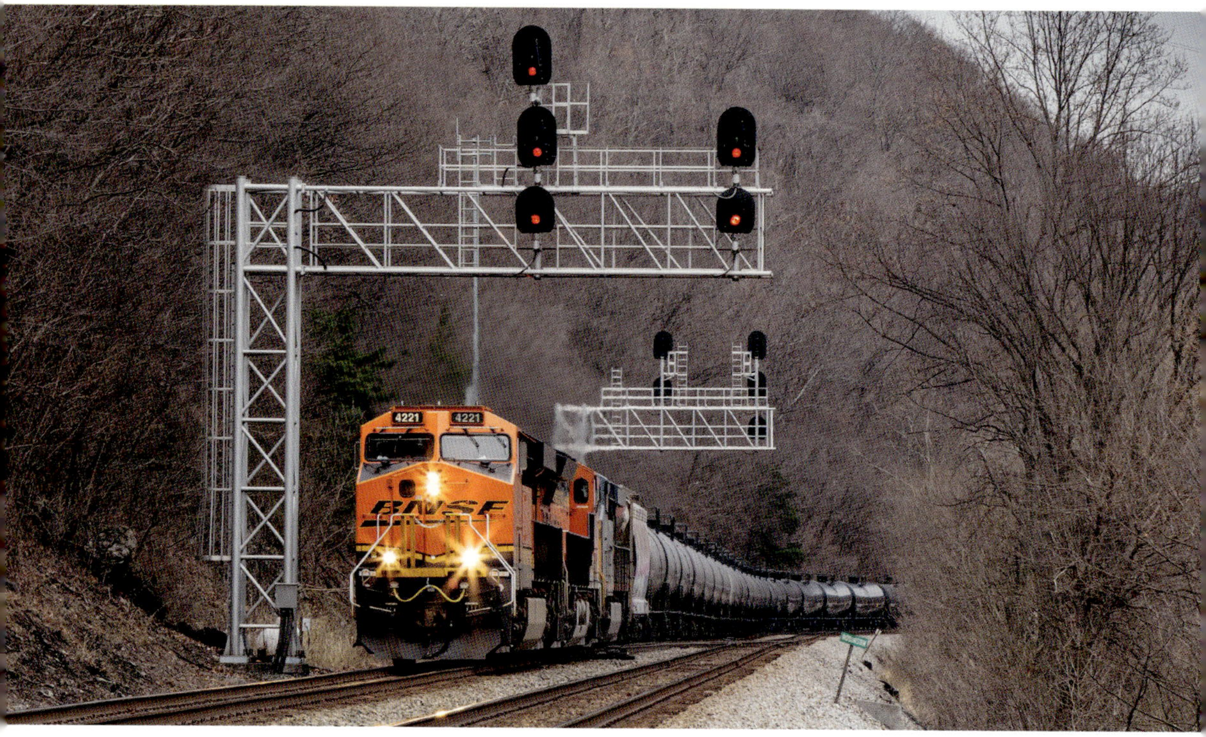
4221
4221
BNSF

Seen from the public boat ramp in Rich Creek, intermodal NS 217 rushes west alongside the New River. The westbound signal of control point Robinson is just ahead of it on this overcast July morning.

*Opposite above:* Continuing west along the line we come to Rich Creek, Virginia. This view is from the Robertson Mountain Road crossing off of Lurich Road. With the signal at control point Robinson in its windshield, the Illinois Terminal heritage SD70ACe leads a domestic coal hopper train west toward Bluefield at the end of a spring day.

*Opposite below:* An overcast January afternoon finds a loaded ethanol train, NS 6D4, thundering east through the signals at control point Robinson with a BNSF leader. Control point Robinson is milepost 336.4.

Mid-morning light shines down on AC44C6M 4001 as it leads an extra 234, NS i34, east through Rich Creek, VA.

*Opposite above:* A dark and overcast September morning finds westbound intermodal NS 233 rolling through Rich Creek, Virginia. To my right is the Lurich Road crossing. Lurich Road will continue west alongside the rails to Glen Lyn, Virginia.

*Opposite below:* In some gorgeous morning light, loaded ethanol train NS 6D4 speeds east through Rich Creek at the Lurich Road crossing. The Canadian Pacific SD70ACU rebuild commemorating the livery of the Canadian and American fighter jets with its two-tone gray paint scheme leads the train.

On a dark November morning, intermodal NS 217 heads west through Glen Lyn, Virginia, with a BNSF trio on point. The leader is an EMD SD75M, which is quite rare to see through here. The close track in the photo was once used for servicing the power plant just ahead with coal hoppers.

Under cloudy skies, hopper train NS 821 heads west along Lurich Road with an ES44AC in the lead. No. 8184 was built in March 2017 for Norfolk Southern and is the highest numbered ES44AC on the roster.

The now closed power plant in Glen Lyn, Virginia looms in the background on an overcast and cold February day as a loaded grain train rumbles east with an ES44AC leader. Lurich Road is in the foreground.

The original Norfolk Southern heritage unit is in the lead of manifest NS 19M as it passes by the now closed American Electric Power plant in Glen Lyn, Virginia. When the power plant closed on December 31, 2014, a pair of switching engines from the facility were saved and now reside at the Virginia Museum of Transportation in Roanoke.

Day turns to night as manifest NS 352 heads east under the signals at control point Glen Lyn in Glen Lyn, Virginia, with a pair of AC44C6Ms. Glen Lyn Baptist Church can be seen on the hill.

Loaded eastbound grain train NS 54W ducks under the signals at Glen Lyn with an AC44C6M on point. Control point Glen Lyn is milepost N340.5 and is the last town in Virginia before crossing over into West Virginia.

Now we have moved into the mountains of southern West Virginia in Mercer County. Here we see a pair of AC44C6M rebuilds leading a loaded coal train from the coal fields of West Virginia to Norfolk, Virginia for export. The heavy train is seen passing east through Willowton, WV, as it approaches the intermediate signal there.

The Southern heritage and original Norfolk Southern heritage units pull a coal hopper train west over top of Christie Road in Hale's Gap, West Virginia. At the time of this publication, the Southern heritage unit is in storage after a derailment in Pennsylvania that left the unit with quite a bit of damage.

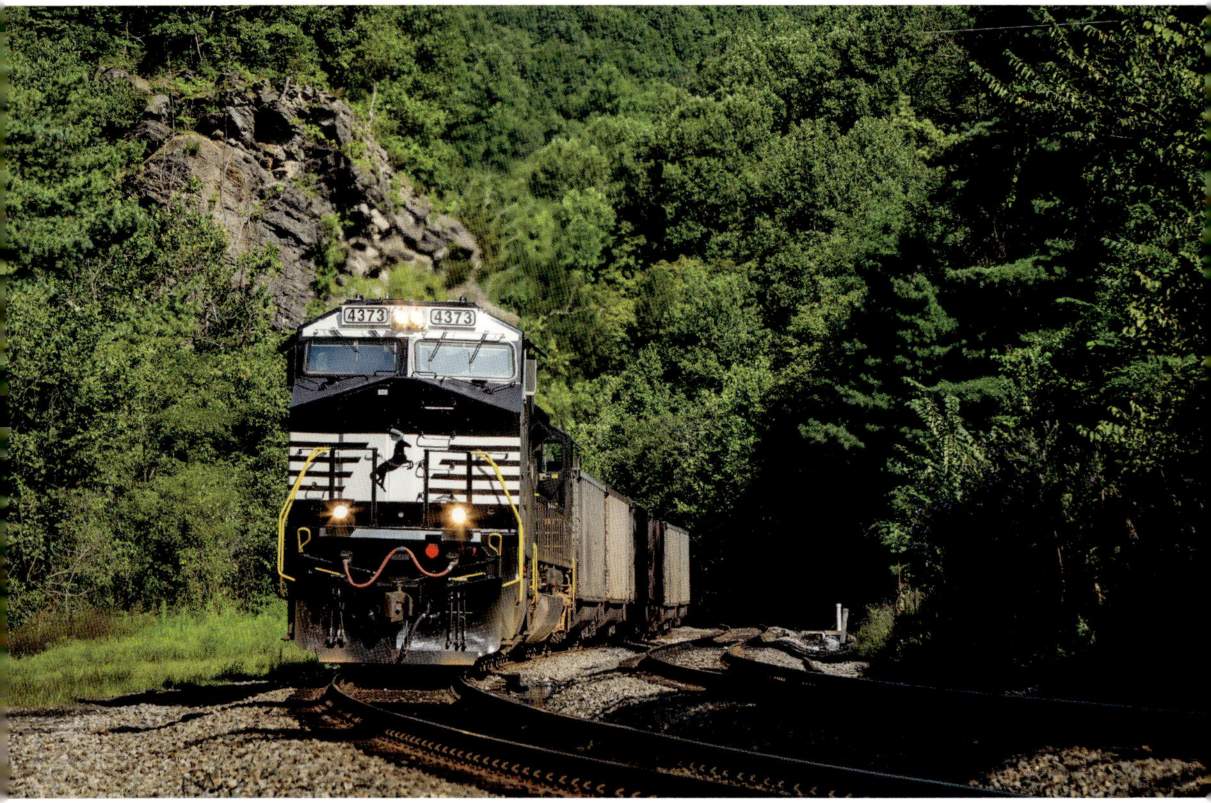
4373
4373

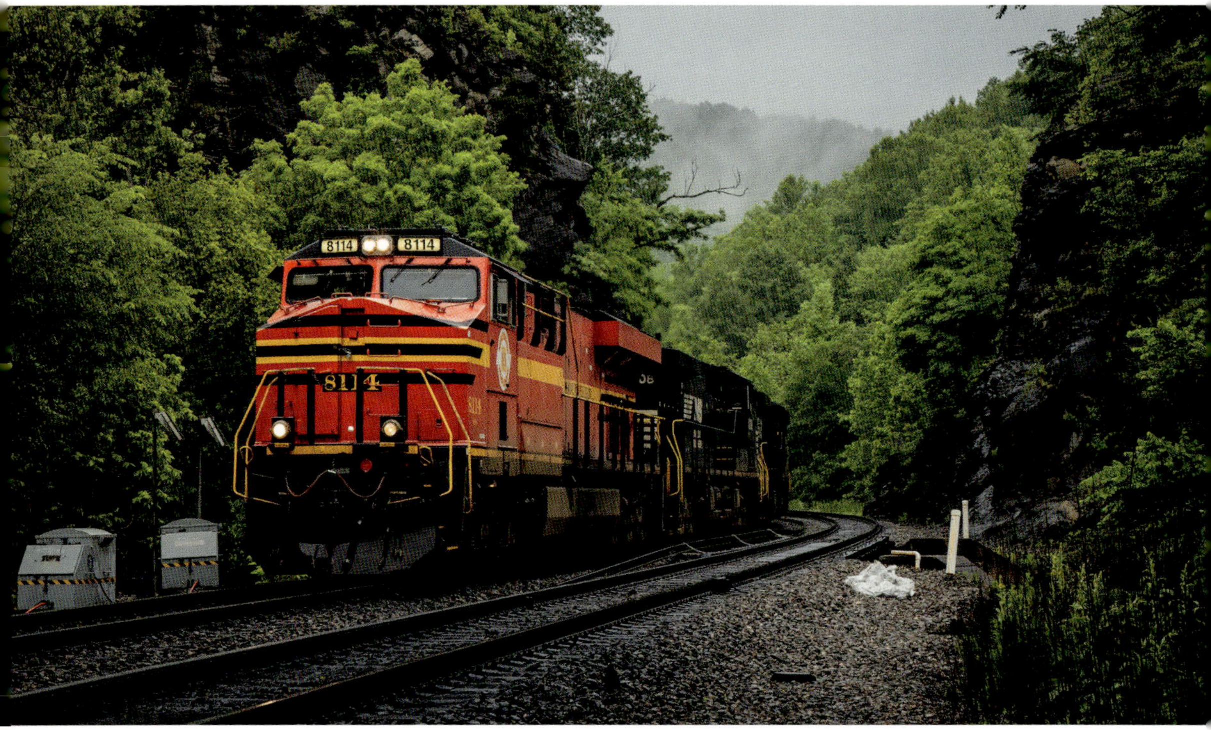
8114
8114
8114

Hopper train NS 821 rolls west toward Bluefield with the first AC44C6M, no. 4000, in the lead at Hale's Gap.

*Opposite above:* Seen from track level at Hale's Gap, a brand new AC44C6M rebuild is on the head of empty export hopper train NS 815 as it thunders west through Hale's Gap, West Virginia. AC44C6M 4373 was rebuilt from Dash 9 9073 in August 2020, which is the same month and year as this photo.

*Opposite below:* ES44AC no. 8114, the original Norfolk Southern heritage unit, leads manifest NS 19M west between the rock walls at Hale's Gap. Rail greasers can be seen to the left of the lead unit.

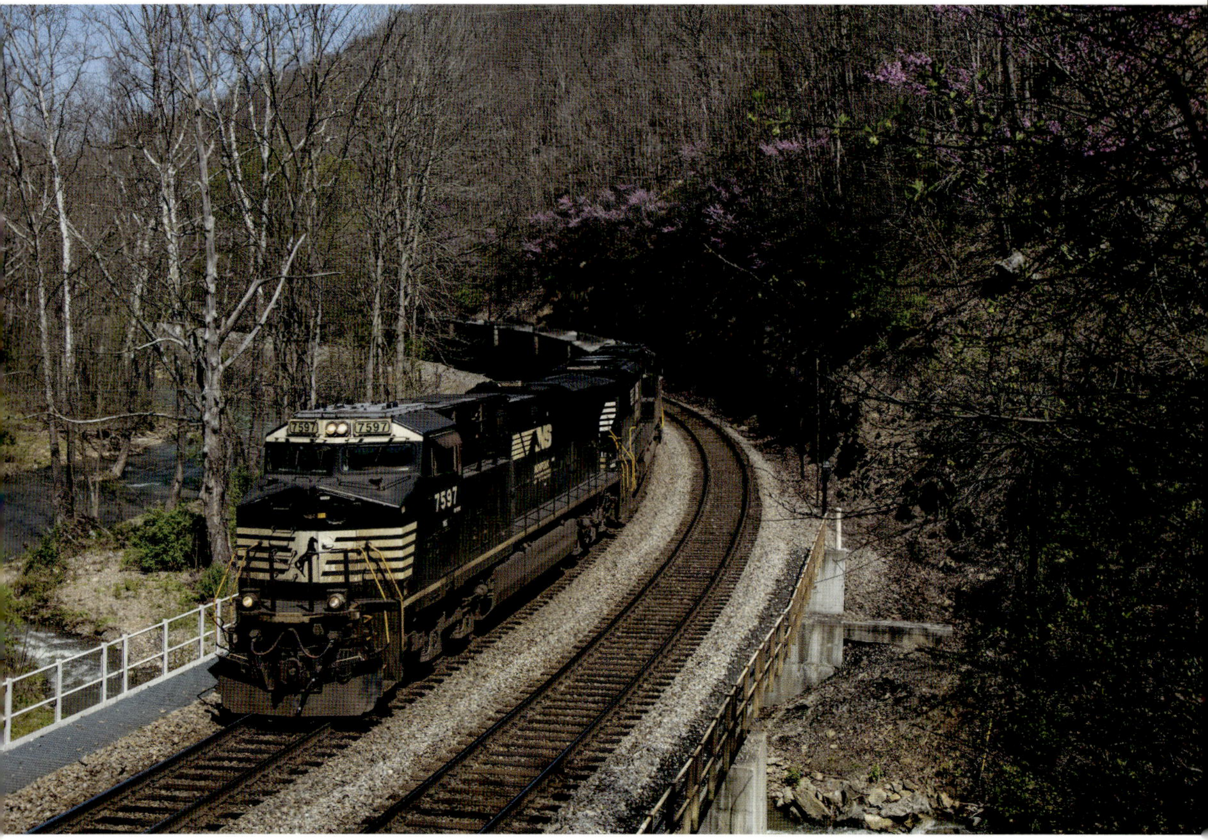
7597
7597
NS
7597

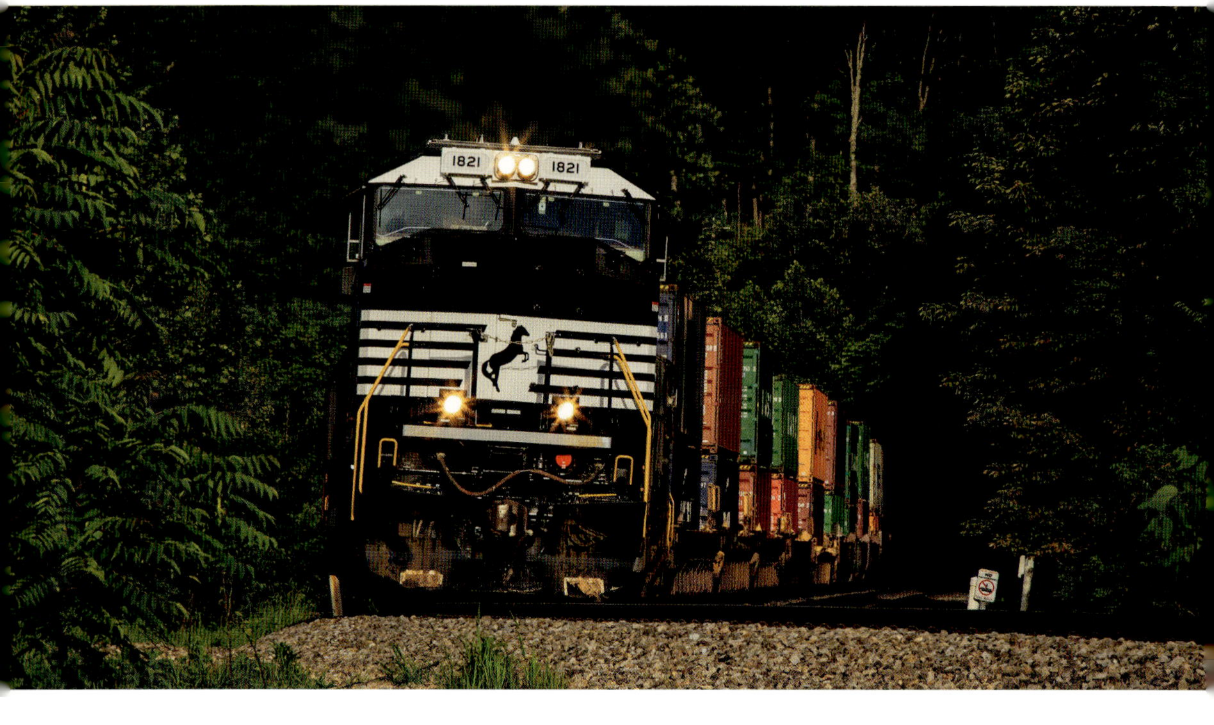
1821
1821

The Nickel Plate Road heritage unit heads up hopper train NS 741 as it rumbles into the tiny town of Oakvale, West Virginia. In the background, you can see a trestle on the Norfolk Southern Princeton Deepwater District. The line starts in Kellysville, WV, and goes into Princeton, WV, to a rock quarry. The line used to go much further into West Virginia, but was recently closed due to inactivity.

*Opposite above:* Now moving west into Oakvale, West Virginia, we see hopper train NS 815 rounding a curve covered in blooming Red Bud trees. This trestle is over the top of Morgan Road and the East River can be seen to the left.

*Opposite below:* Shiny SD70ACC 1821 leads westbound intermodal NS 29G through the S-curve in Oakvale, West Virginia, just before town. No. 1821 was rebuilt from standard cab SD70 no. 2525 in 2019. 29G is a daily intermodal that operates from Norfolk, Virginia, to Rickenbacker Yard in Columbus, Ohio.

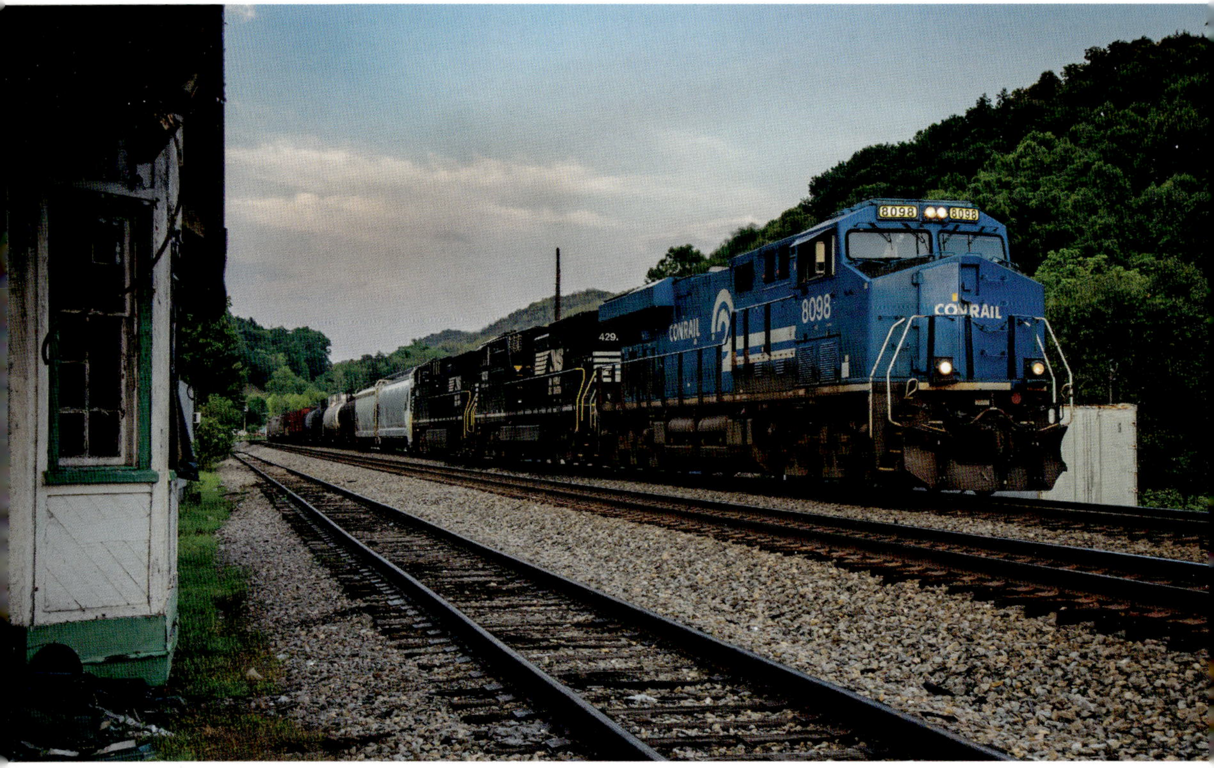
8098
8098
8098
CONRAIL
CONRAIL

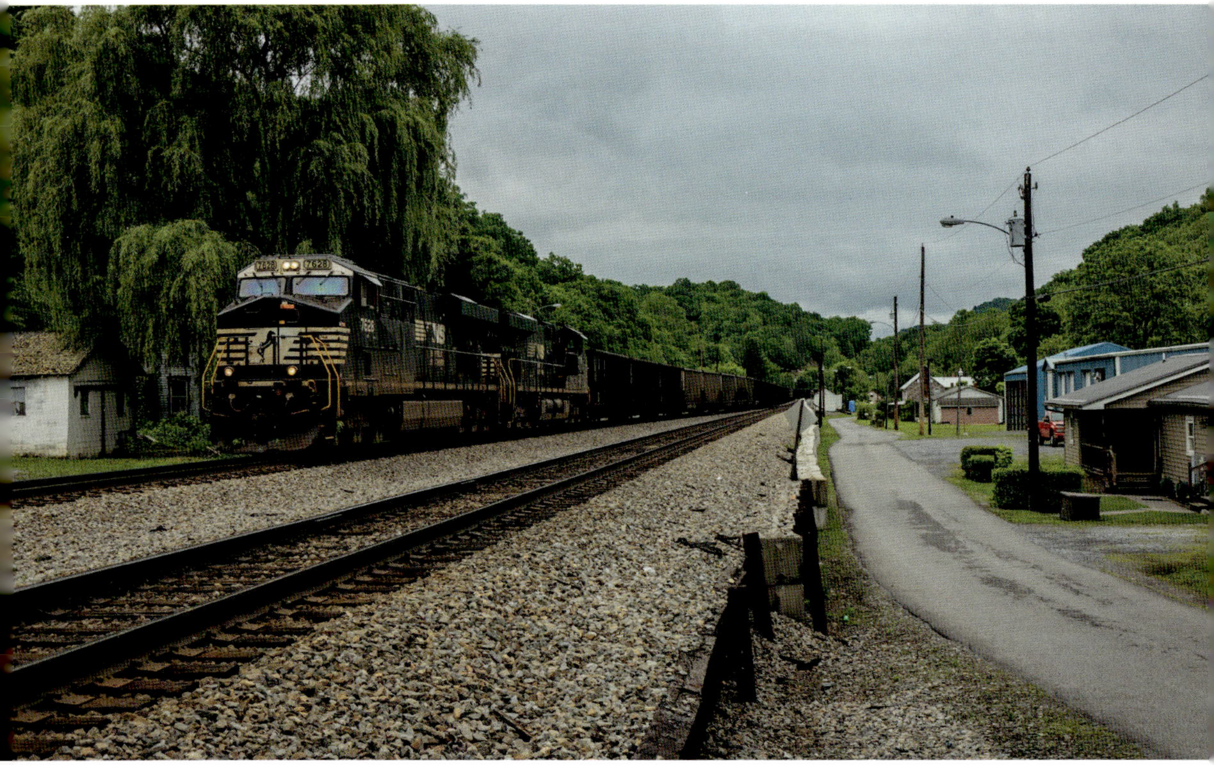

Loaded ethanol train NS 6D4 rolls east over the Ingleside Road crossing in Oakvale, West Virginia. A BNSF Dash 9 in warbonnet paint leads the charge as day turns into night. This "warbonnet" scheme first debuted on the Santa Fe Railroad and was used on their streamlined diesels of the 1930s. The paint scheme was dropped at the end of passenger service in 1971. It was later used again in the late 1980s and when a merger formed the current day Burlington Northern Santa Fe, but has since been stopped for good.

*Opposite above:* Westbound manifest NS 19M rushes by the Oakvale depot with the Conrail heritage unit in charge. The depot is quite dilapidated at the time of this photo. Conrail officially began on April 1, 1976, after congress passed the Railroad Revitalization and Regulatory Reform Act. Conrail was short for the Consolidated Rail Corporation.

*Opposite below:* A wider view of Oakvale, WV, shows an empty hopper train rolling west. Oakvale was incorporated in 1907 and was named for the large Oak trees growing in the area. Oakvale was originally named Frenchville in honor of the family that first settled here; their last name was French.

The Norfolk Southern OCS (Officer Car Special) speeds west through Oakvale on main two. These F units have since been sold to two separate railroads and now Norfolk Southern uses a pair of EMD SD60Es for the train's power.

*Opposite above:* A telephoto view from the road crossing in Oakvale, WV, shows intermodal NS 217 rushing west by the depot with an old logo Dash 9 in the lead. No. 9588 was built in January 2001 for Norfolk Southern.

*Opposite below:* The blue-nosed AC44C6M no. 4001 is seen approaching the end of the straight stretch at control point Oakvale with export hopper train NS 815. Control point Oakvale is milepost N347.5.

High above Ingleside Road, a loaded grain train rushes east through Oakvale, WV, with a pair of Dash 9s for power.

*Opposite above:* An EMD SD70IAC and a GE Dash 9 pull empty ethanol train NS 65W west by a private road crossing just off of Ingleside Road in Oakvale, West Virginia. SD70IACs were numbered 1175 to 1224 and had slight changes from the SD70ACes.

*Opposite below:* Canadian Pacific SD70ACU 7023 leads NS 6D4 east by a private road crossing off of Ingleside Road through Oakvale.

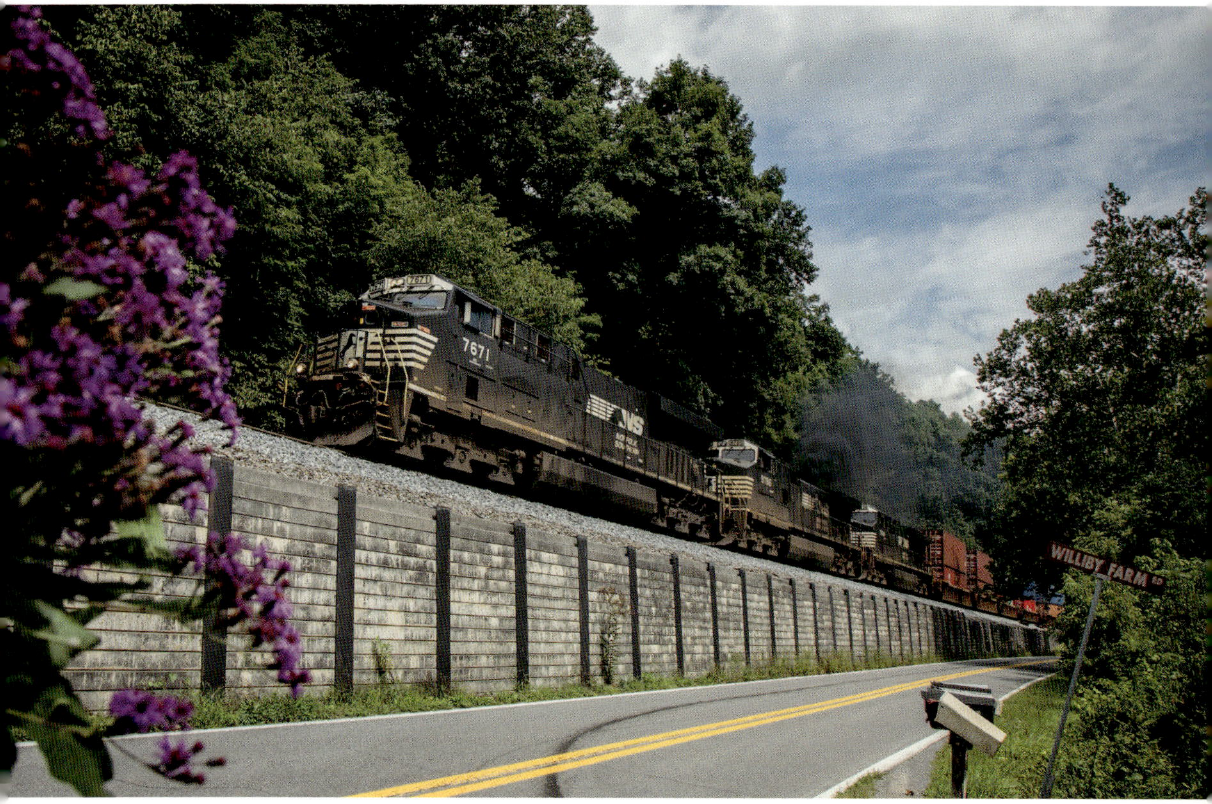
7671
WILLIBY FARM

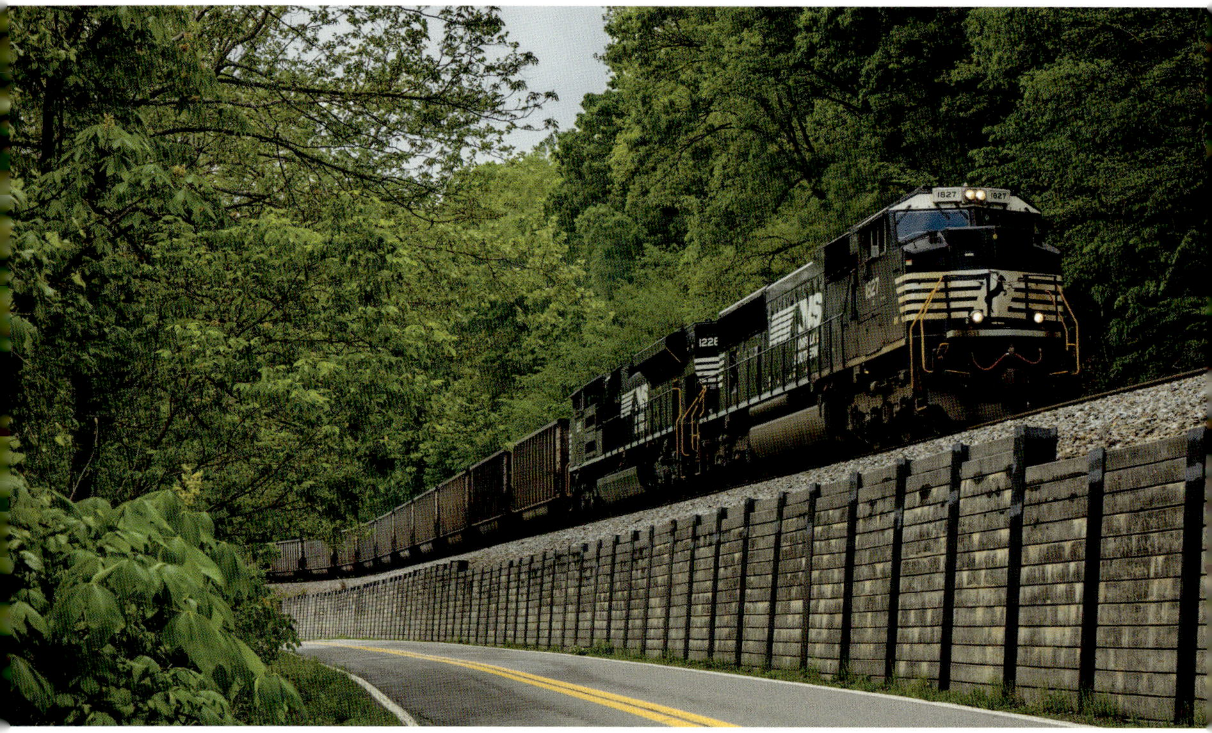
1827
1228

AC44C6M 4001 is seen leading an extra 234, NS i34, east through Hardy, West Virginia, on a warm September afternoon. i34 terminates in Norfolk, Virginia.

*Opposite above:* Just up Ingleside Road on the outskirts of Oakvale, the line sits above the road with a barrier wall in place which makes for an interesting view. On this day an ES44DC and smokey Dash 9 pull intermodal NS 217 west along the wall toward their final destination of Chicago, Illinois.

*Opposite below:* SD70ACC rebuild no. 1824 pulls loaded coal train NS 822 east alongside Ingleside Road. No. 1824 is ex-standard cab SD70 no. 2555. A brand new EMD SD70IAC rides second.

Under clear blue skies, intermodal NS 217 heads west right outside of Oakvale, West Virginia, with a Burlington Northern Santa Fe trio. BNSF 298 is an EMD SD75M that was originally in warbonnet paint.

A pair of GEs pull loaded coal train NS 820 east through a snowy Hardy, West Virginia, as it approaches the Peggy Branch Road crossing. In the background, the tail end of westbound manifest NS 353 is visible.

NS 65W rushes west toward the Peggy Branch Road crossing in Hardy, West Virginia. NS 56W is an as needed empty ethanol train that operates from Selma, North Carolina, to Chicago, Illinois.

A pair of Union Pacific GE products lead an empty ethanol train west over Ingleside Road in Hardy, WV. On the other side of this bridge, near the waterfall, once stood a mill that was built before the railroad was built.

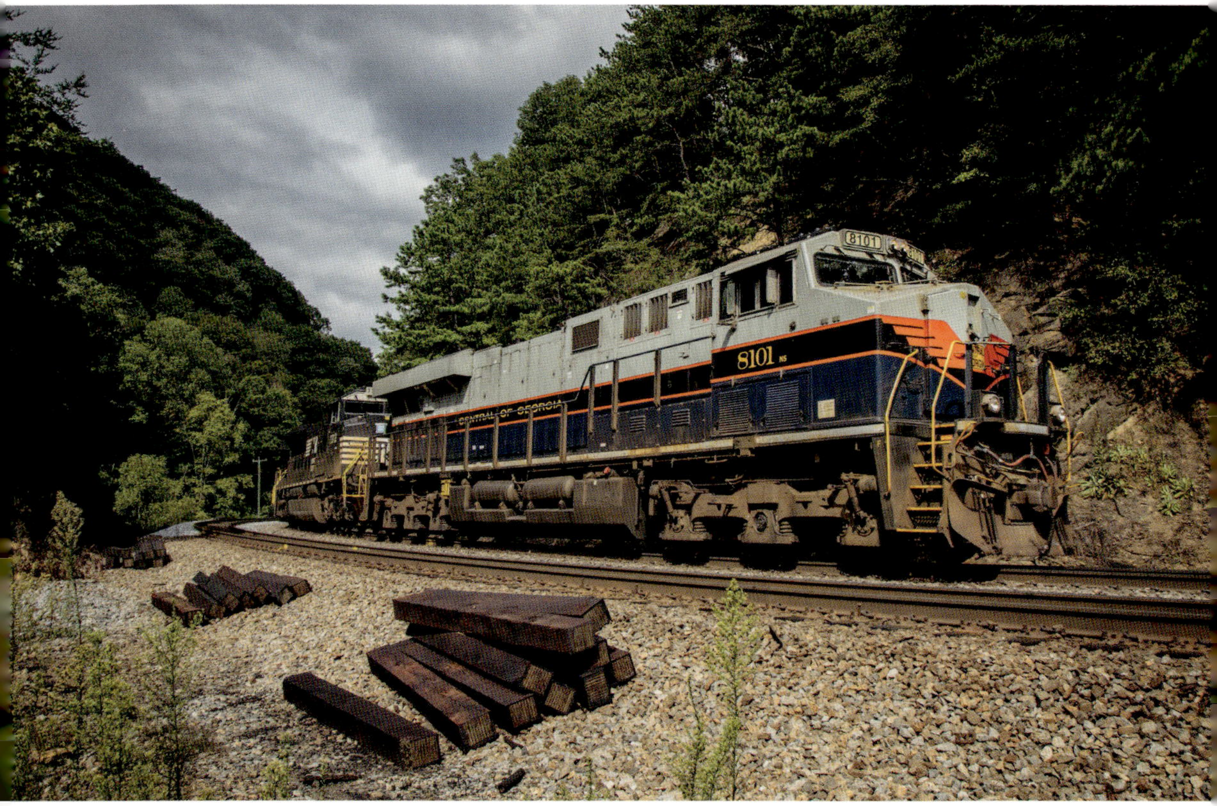
8101
8101
CENTRAL OF GEORGIA

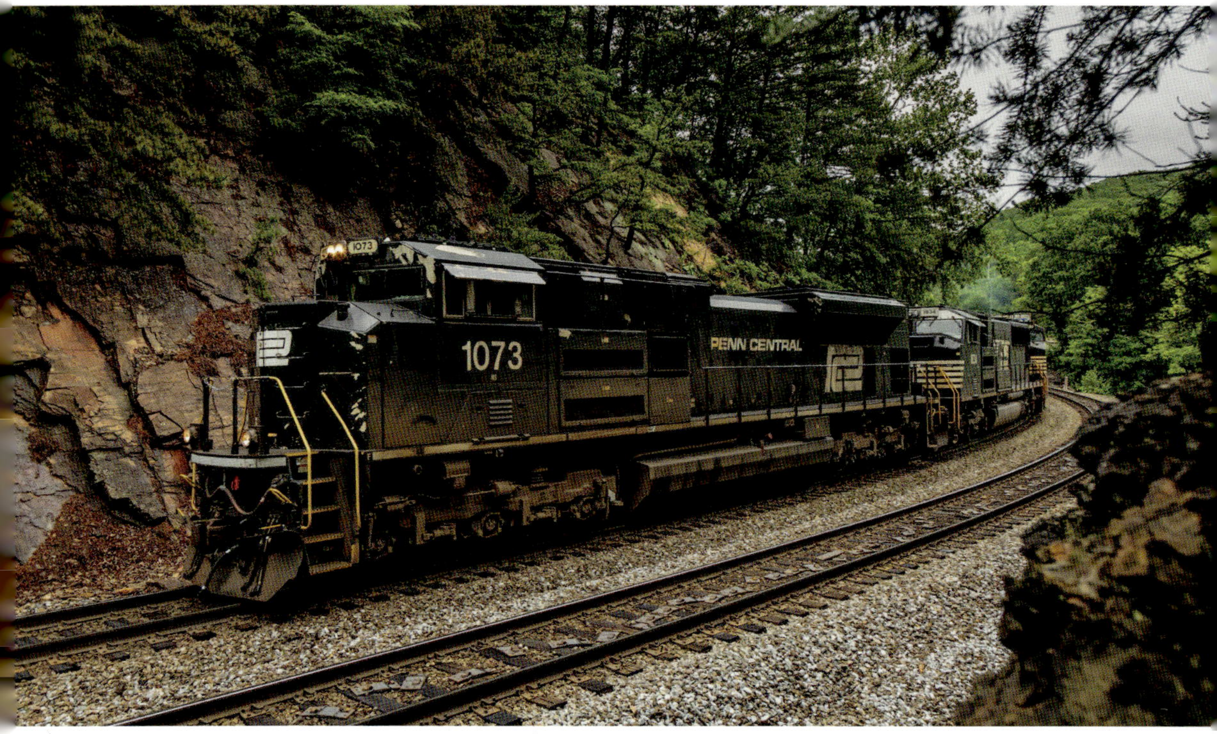
1073
1073
PENN CENTRAL

From the same vantage point as the last photo, just turned the opposite direction, a coal train hurries east with an SD70ACC leading. The western end of the Christiansburg district offers many different photo angles with all of the twists and turns through the mountains.

*Opposite above:* Looking west at the same spot as the last photo, eastbound manifest NS 18M rushes along the Christiansburg district with the Central of Georgia ES44AC heritage unit in the lead. Ties lay along the mainline as soon there will be a major replacement project.

*Opposite below:* Twists and turns are abundant here on the western end of the Christiansburg district as the line meanders toward Bluefield. Here the Penn Central heritage unit negotiates one of the tight curves in Hardy as it leads train 217.

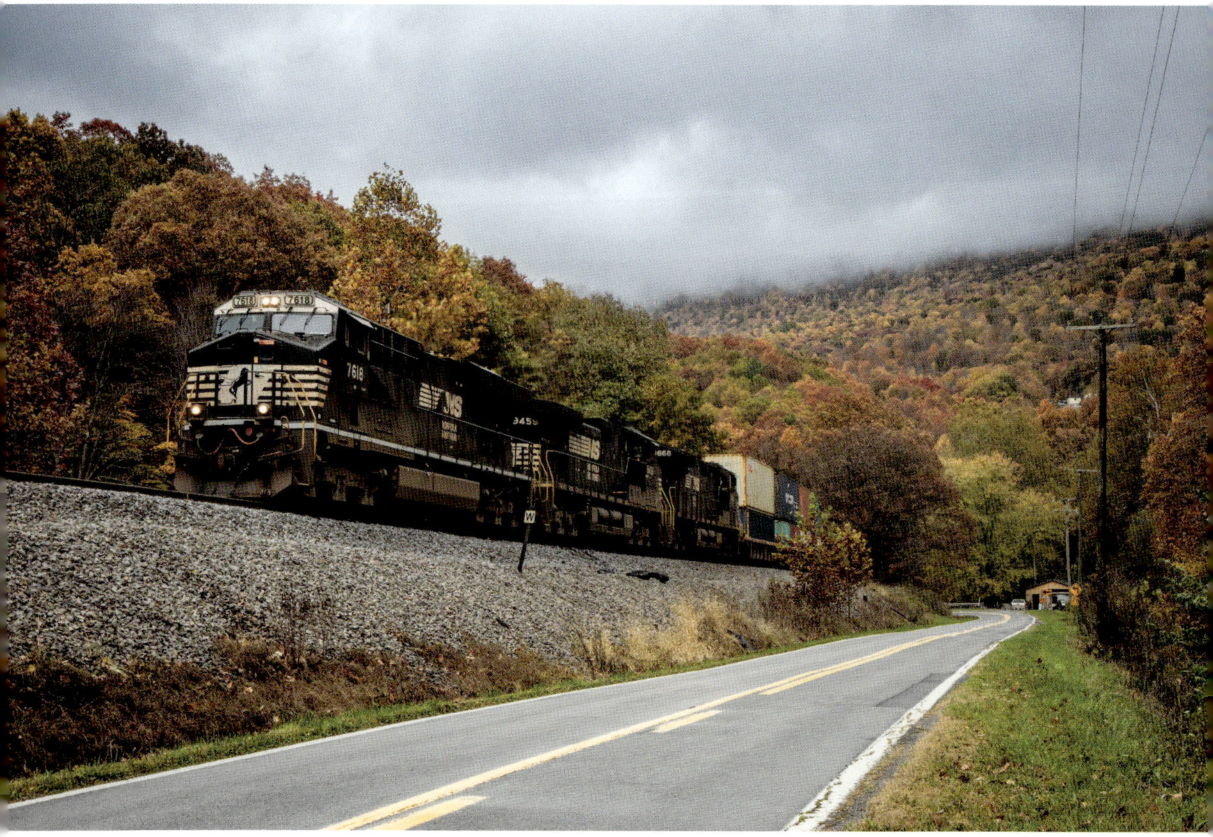

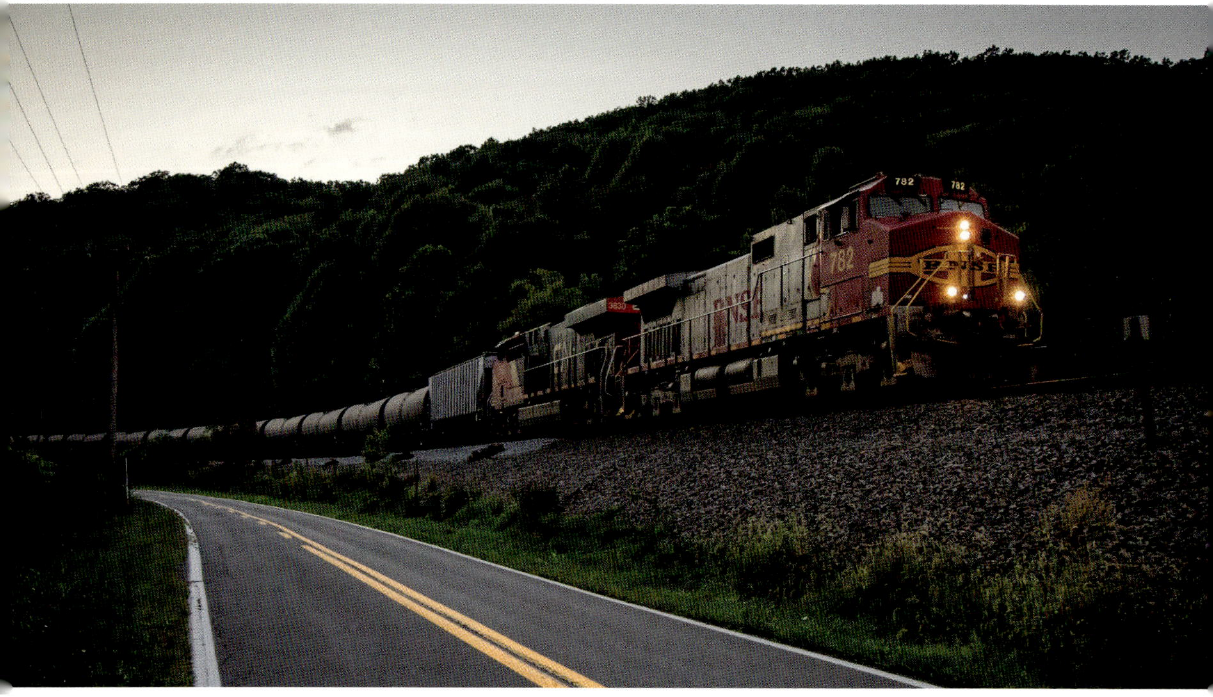

A Dash 9 lugs intermodal NS 233 through the S-curve in Hardy, WV, on a fall morning. This train had grain hoppers placed in front of the relatively short intermodal. No. 9524 was built in May 2000 for Norfolk Southern. Not long after this photo, this unit was placed into storage in preparations for rebuilding into an AC44C6M.

*Opposite above:* Fall colors are showing in the mountains of Southern West Virginia as NS 217 rockets west along Ingleside Road with an ES44DC leader. Ingleside Road follows the mainline from Oakvale to Ada, West Virginia.

*Opposite below:* With just minutes of daylight left, loaded ethanol train NS 6D4 speeds east through Ingleside, West Virginia. Running alongside Ingleside Road, the BNSF Dash 9 in warbonnet paint powers the train toward Roanoke, Virginia where the power was pulled and black NS units continued the rest of the trip to Selma, North Carolina.

Negotiating an S-curve in Hardy, West Virginia, is manifest NS 393. The Wabash SD70ACe heritage unit is on point of the train on a sunny March afternoon. The Wabash Railroad operated from 1837 to 1964 when it was merged into the Norfolk and Western Railway. No. 393 is a daily manifest/autorack train from Winston Salem, NC, to Portsmouth, OH.

A Canadian Pacific GE leads eastbound loaded ethanol train NS 6D4 east through Hardy, WV. 6D4 is notorious for having foreign power. These trains originate in Chicago, Illinois, and terminate in Selma, North Carolina.

Canadian Pacific SD70ACU 7023 leads eastbound loaded ethanol train NS 6D4 around a curve in Ingleside, West Virginia.

Late evening light shines down on the nose of AC44C6M 4208 as it pulls manifest NS 393 west toward the Cox's Cutoff intermediate signal in Ingleside, West Virginia.

The Central of Georgia heritage unit leans into a curve at Cox's Cutoff in Ingleside, West Virginia, with eastbound manifest NS 18M.

Seen from above, loaded coal train NS 820 thunders through Ingleside, WV. Ingleside Road and the East River can be seen paralleling the main line in the small community.

On a fall morning, NS 217 rushes west through Ingleside, WV. An ES44DC leads the charge toward Bluefield for a fresh crew. 7618 was built in February 2007.

Manifest NS 19M negotiates a little bit of an S-curve as it heads toward control point Blake. A very clean BNSF EMD SD70ACe leads the long freight train.

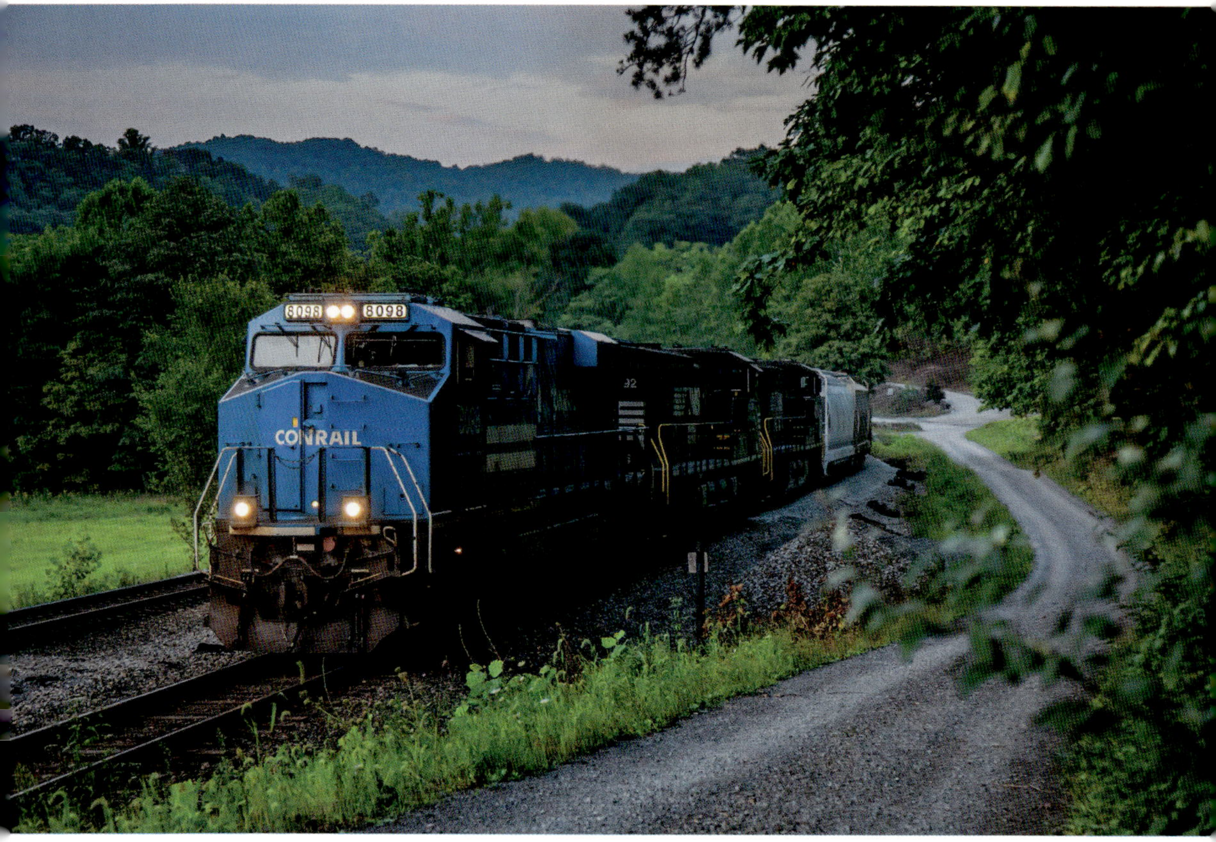
8098
8098
CONRAIL

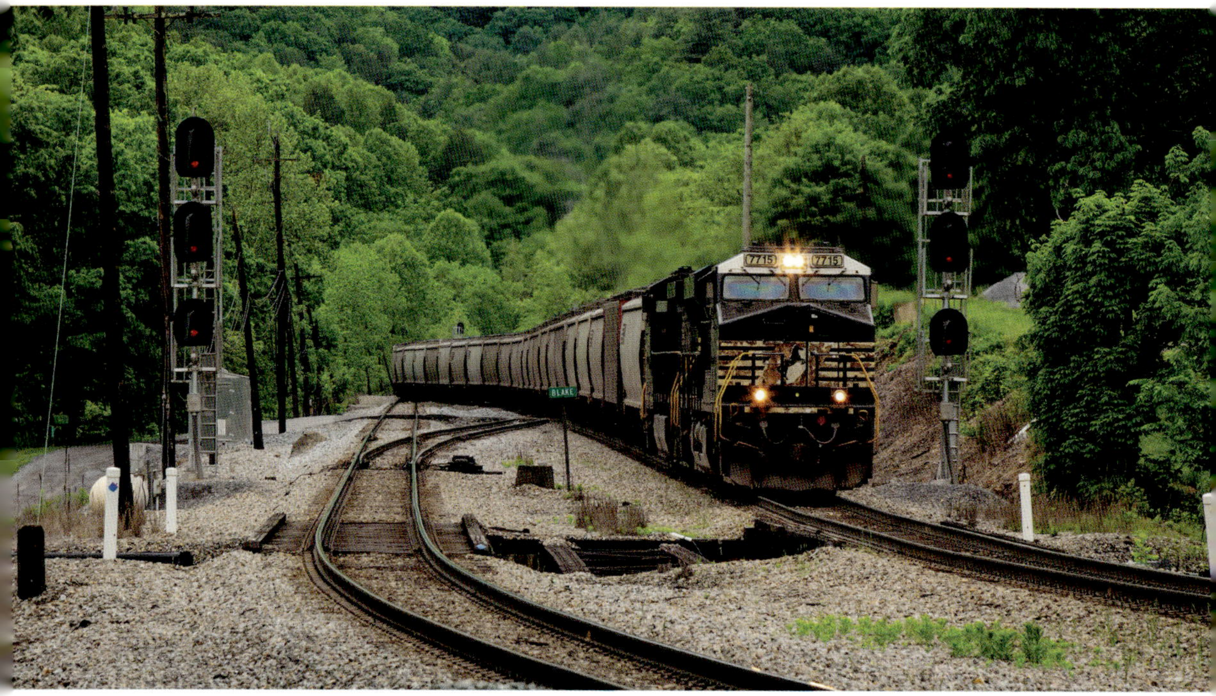
7715
7715
BLAKE

Westbound domestic coal hopper train NS 741 rounds the curve leading to Ada, WV, on a beautiful summer afternoon with the Nickel Plate Road heritage unit in charge.

*Opposite above:* Day turns to night as manifest NS 393 rolls toward control point Blake alongside the gravel Blake Road with the Conrail heritage unit on point.

*Opposite below:* A rusty-nosed ES44DC leads loaded grain train NS 40N as it switches from main one to main two and splits the signals of Blake. Control point Blake is milepost N354.3.

The central of Georgia heritage unit is seen rushing east through Ada, West Virginia, with manifest NS 18M on a cloudy September morning. This view is from the TNG Lane crossing.

*Opposite above:* A Union Pacific ES44AC is in charge of westbound intermodal NS 217 as it makes its way through a cold Ada, WV.

*Opposite below:* A Canadian Pacific GE AC44CW is on the head end of a heavy loaded ethanol train as it rolls through the intermediate signal in Ada.

A pair of westbounds race into Bluefield on a sunny early spring afternoon. Main two is occupied by an empty export coal hopper train with an ES44DC in the lead. A rail greaser is seen next to the second unit.

*Opposite above:* On Valentine's Day 2021, empty grain train NS 51N heads west through Ada, West Virginia, with a specially painted SD60E to honor first responders. The unit was aptly numbered 9-1-1. Norfolk Southern SD60Es were rebuilt from standard cab SD60s, and most are relegated to local services now.

*Opposite below:* A DC (direct current) GEVO pulls a loaded ethanol train east under the intermediate signal in Ada. It is seen here approaching the Ada Road crossing.

BNSF
4221
BNSF
3823
BNSF

An empty grain train is surrounded by fall colors as it snakes its way into Bluefield yard after traversing the Christiansburg district. All empty grain and ethanol trains go through a thorough inspection in the yard before continuing onto the Pocahontas district.

*Opposite above:* We have finally reached Bluefield, West Virginia, and we are almost to Bluefield yard, marking the end of the Christiansburg district. Here intermodal NS 217 rides along a rock wall as it slowly makes its way toward the yard with the Jersey Central heritage SD70ACe in the lead. This vantage point is from the shoulder of a bridge on US 460.

*Opposite below:* A pair of BNSF GEs and a CSX GE unit pull loaded ethanol train NS 6D4 out of Bluefield yard on a cold and overcast January afternoon.

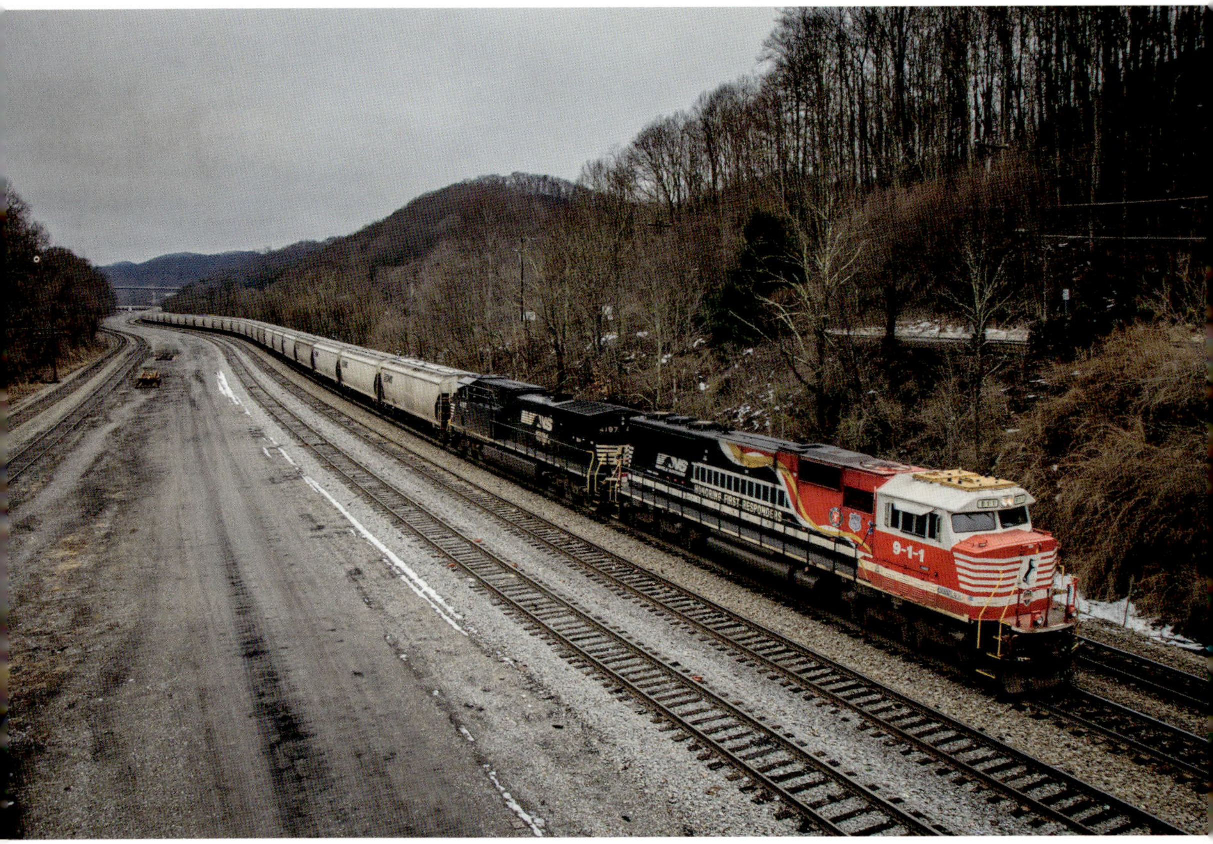
9-1-1
HONORING FIRST RESPONDERS

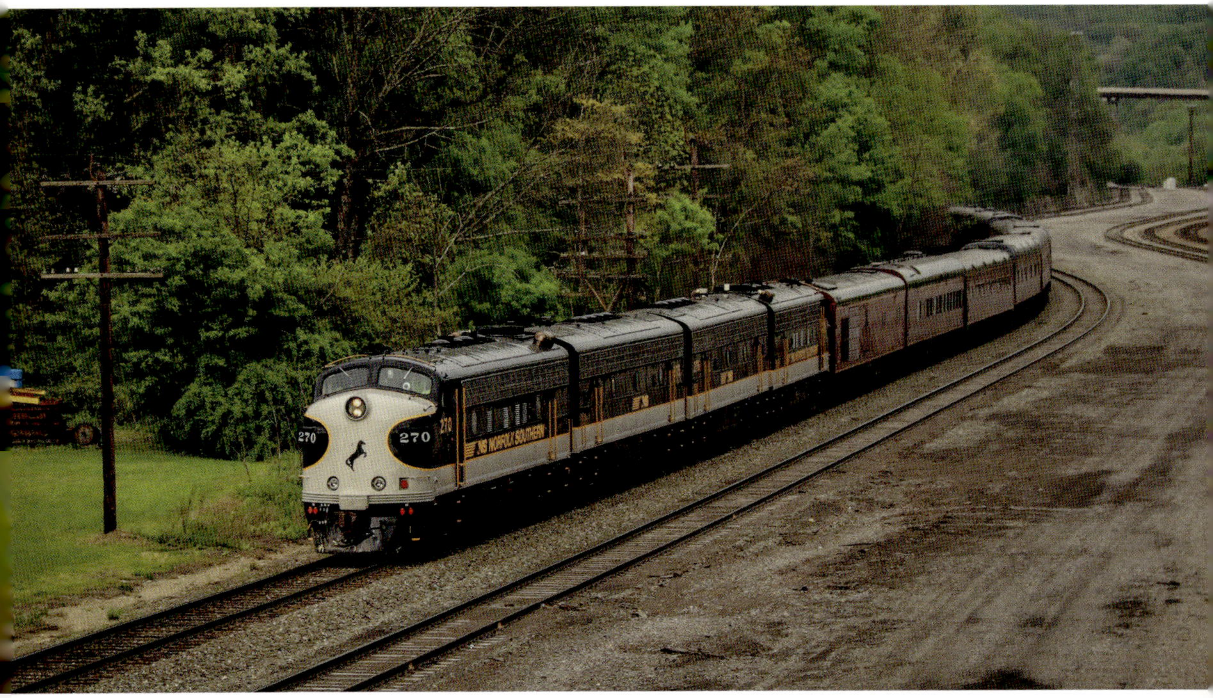
270
270
NORFOLK SOUTHERN

Manifest NS 194 rolls east out of Bluefield yard as it heads out to traverse the Christiansburg district. A trio of GE products lead the train as it goes by control point RD, which can be seen in the background. Control point RD is milepost N360.5.

*Opposite above:* The honoring first responders SD60E leads empty grain train NS 51N into Bluefield yard.

*Opposite below:* The Norfolk Southern officer car special snakes its way into Bluefield yard for a quick crew change before continuing west onto the Pocahontas district.

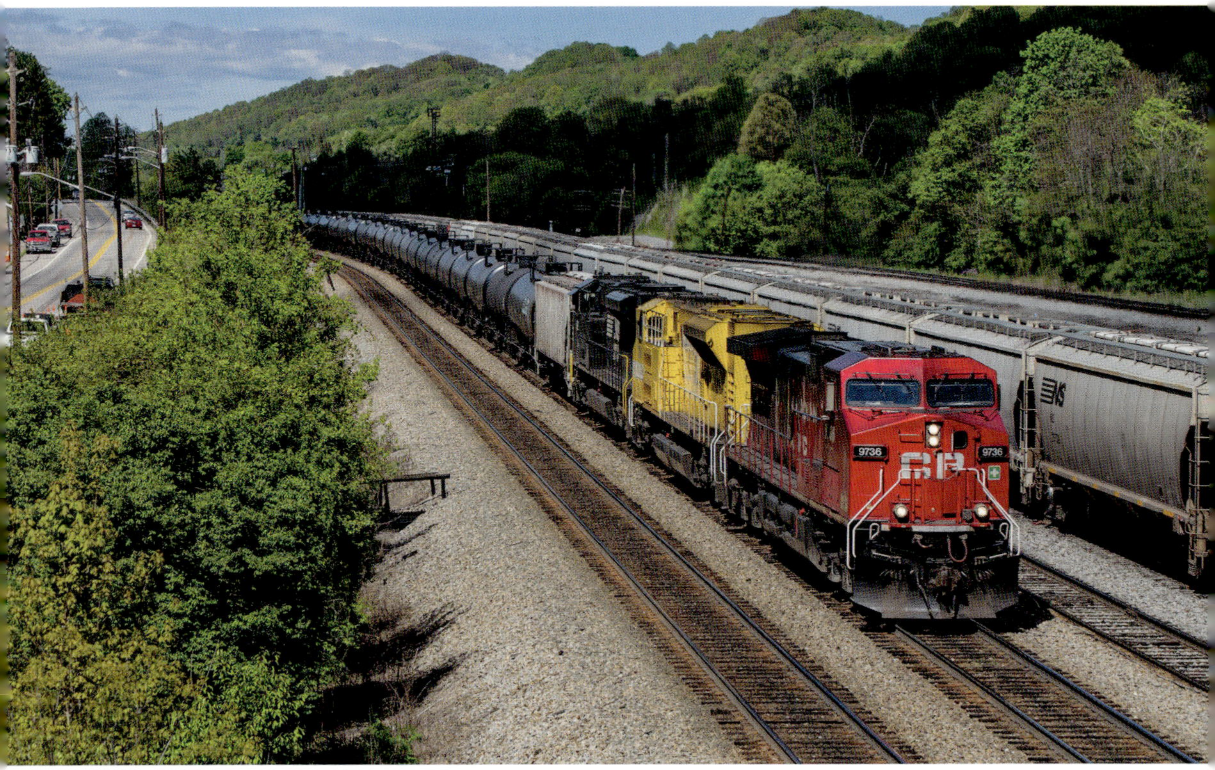
9736
CP
9736
NS

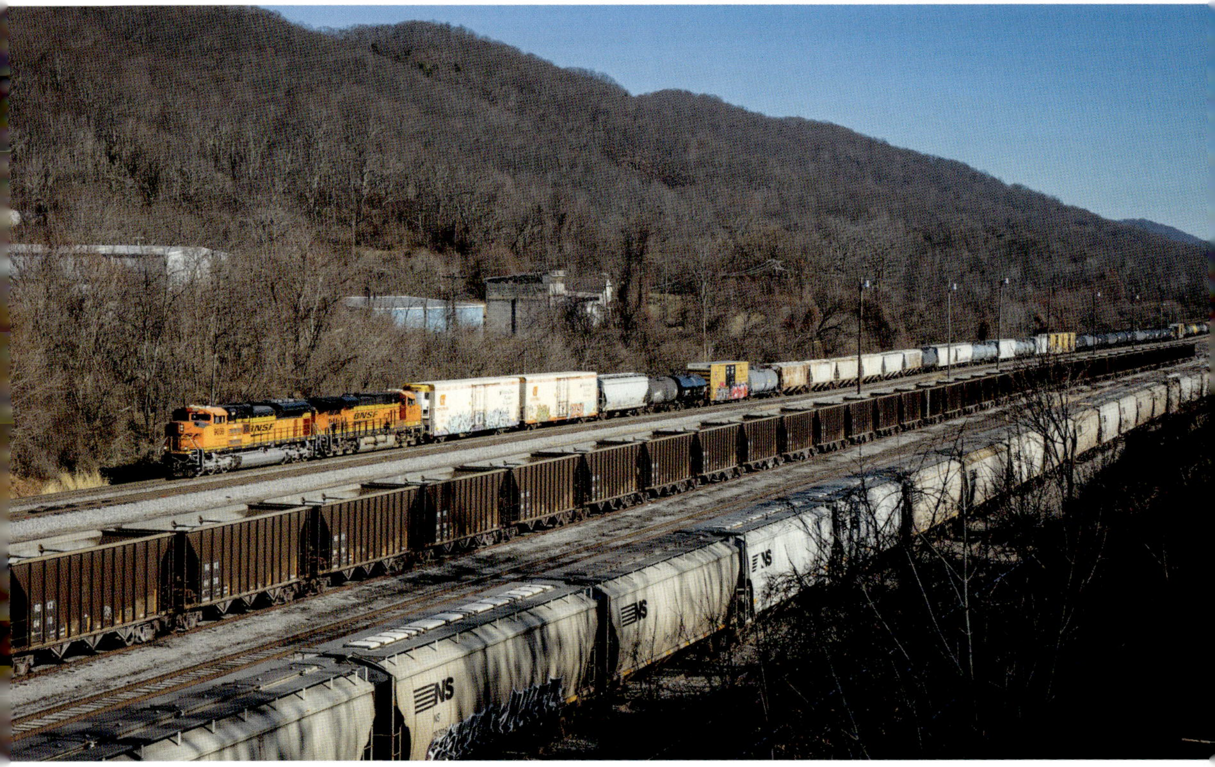
BNSF
BNSF
NS
NS
NS

With the original Norfolk Southern heritage unit in the lead, NS 19M has made its way into Bluefield yard and is just a few hundred yards away from a crew change. This view is from the Grant Street bridge, which has since been demolished. One of the yard tracks has also been taken out, as evident on the left side of this photo.

*Opposite above:* A Canadian Pacific GE slowly pulls out of Bluefield yard and is about to head out onto the Christiansburg district with a loaded ethanol train. A bustling Princeton Avenue can be seen to the left.

*Opposite below:* Manifest NS 19M has made it into Bluefield yard with a BNSF duo as it runs the Virginia Pull In track toward a crew change. Years ago, when coal was still king, this yard would be full of trains, primarily coal trains. Now, there are not nearly as many trains through here a day as there used to be.

The sun is setting on a warm November day in Bluefield yard. Hopper train NS 821 is getting refueled and a fresh crew will jump aboard before continuing west toward the coal fields of West Virginia. The Norfolk and Western built coaling tower still stands tall today as it sees trains come and go all day and night. Norfolk Southern's famed Pocahontas district is just ahead and the mountains and tough grades will continue for trains. With this, we end our journey across the 100-mile Norfolk Southern Christiansburg district.

# Bibliography

"Cambria Depot." Virginia Center for Civil War Studies, civilwar.vt.edu/cambria-depot/.

"Conrail," Wikipedia, Wikimedia Foundation, June 28, 2022, en.wikipedia.org/wiki/Conrail

"Lehigh Valley Railroad." Wikipedia, Wikimedia Foundation, June 26, 2022, en.m.wikipedia.org/wiki/Lehigh_Valley_Railroad.

"Heartland Corridor," Wikipedia, Wikimedia Foundation, May 5, 2018, en.m.wikipedia.org/wiki/Heartland_Corridor.

"Illinois Terminal Railroad," Wikipedia, Wikimedia Foundation, March 13, 2022, en.m.wikipedia.org/wiki/Illinois_Terminal_Railroad#:~:text=The%20Illinois%20Terminal%20Railroad%20Company,Illinois%20from%201896%20to%201956.

"Lehigh Valley Railroad," Wikipedia, Wikimedia Foundation, June 26, 2022, en.m.wikipedia.org/wiki/Lehigh_Valley_Railroad.

"Monongahela Railway," Wikipedia, Wikimedia Foundation, July 15, 2022, en.m.wikipedia.org/wiki/Monongahela_Railway#:~:text=The%20Monongahela%20Railroad%20was%20a,first%20meeting%20in%20January%201901.

"Narrows, Virginia." Wikipedia, Wikimedia Foundation, February 28, 2022, en.m.wikipedia.org/wiki/Narrows,_Virginia.

"Norfolk and Western Railway," Wikipedia, Wikimedia Foundation, July 13, 2022, en.wikipedia.org/wiki/Norfolk_and_Western_Railway.

"NS Train Symbols," NS Train Symbols: RailroadfanWiki, railroadfan.com/wiki/index.php/NS_Train_Symbols.

"NSDash9.Com Home Page," NSDash9.Com Home Page, nsdash9.com/.

"Oakvale, West Virginia," Wikipedia, Wikimedia Foundation, July 4, 2022, en.m.wikipedia.org/wiki/Oakvale,_West_Virginia#:~:text=Oakvale%20was%20originally%20known%20as,The%20town%20incorporated%20in%201907.

"Reading Company," Wikipedia, Wikimedia Foundation, July 3, 2022, en.m.wikipedia.org/wiki/Reading_Company#:~:text=The%20Port%20Reading%20Railroad%20was,the%20presidency%20of%20Archibald%20A.

"Town of Pearisburg, VA," TOWN OF PEARISBURG, VA, pearisburg.org/community/welcome_to_pearisburg/history.php#:~:text=Pearisburg%20was%20established%20when%20Capt,the%20intersection%20of%20Main%20St.

"Vicker, Virginia," Wikipedia, Wikimedia Foundation, January 27, 2021, en.m.wikipedia.org/wiki/Vicker,_Virginia.

"Virginia and Tennessee Railroad," Wikipedia, Wikimedia Foundation, August 16, 2021, en.wikipedia.org/wiki/Virginia_and_Tennessee_Railroad.

Gordon, E., "Warbonnet: Railfan Favorite," *Tehachapi News*, September 6, 2016, tehachapinews.com/lifestyle/warbonnet-railfan-favorite/article_8d96b97a-7893-51ae-8219-7cfd9b4d0a08.html.